movement
integration

the systemic approach to human movement

movement
integration

the systemic approach to human movement

Martin Lundgren
Linus Johansson

Contributors

Gary Carter, Åsa Åhman, Julian Baker, Cecilia Gustafsson,
Lucas Henriksson, Lena Björnsdotter, Gary Ward, Jerry Hesch

lotus
publishing

Chichester, England

North
Atlantic
Books
Berkeley, California

First published in 2020 by
Lotus Publishing
Apple Tree Cottage, Inlands Road, Nutbourne, Chichester, PO18 8RJ, and
North Atlantic Books
Berkeley, California

Illustrations and Cover Design Linus Johansson and Martin Lundgren
Text Design Medlar Publishing Solutions Pvt Ltd., India
Printed and Bound in India by Replika Press

Movement Integration: The Systemic Approach to Human Movement is sponsored and published by the Society for the Study of Native Arts and Sciences (dba North Atlantic Books), an educational nonprofit based in Berkeley, California, that collaborates with partners to develop cross-cultural perspectives; nurture holistic views of art, science, the humanities, and healing; and seed personal and global transformation by publishing work on the relationship of body, spirit, and nature.

North Atlantic Books' publications are available through most bookstores. For further information, visit our website at www.northatlanticbooks.com or call 800-733-3000.

Medical Disclaimer
The following information is intended for general information purposes only. Individuals should always consult their health care provider before administering any suggestions made in this book. Any application of the material set forth in the following pages is at the reader's discretion and is his or her sole responsibility.

British Library Cataloging-in-Publication Data
A CIP record for this book is available from the British Library
ISBN 978 1 905367 95 5 (Lotus Publishing)
ISBN 978 1 623174 65 1 (North Atlantic Books)

Library of Congress Cataloging-in-Publication Data
Names: Lundgren, Martin, author. | Johansson, Linus, 1981- author.
Title: Movement integration : the systemic approach to human movement / Martin
    Lundgren, Linus Johansson ; contributors, Gary Carter, Åsa Åhman, Julian
    Baker, Cecilia Gustafsson, Lucas Henriksson, Lena Björnsdotter, Gary Ward,
    Jerry Hesch.
Description: Berkeley, California : North Atlantic Books, 2020. | Includes
    bibliographical references. | Summary: "A paradigm-shifting, integrative
    approach to understanding body movement. Drawing on expertise in
    physiotherapy, somatics, sports science, Rolfing, myofascial therapy,
    craniosacral therapy, Pilates, and yoga, the authors assert that a more
    comprehensive understanding of movement is key to restoring the body's
    natural ability to move fluidly and painlessly"-- Provided by publisher.
Identifiers: LCCN 2019025126 (print) | LCCN 2019025127 (ebook) |
    ISBN 9781623174651 (paperback) | ISBN 9781623174668 (ebook)
Subjects: LCSH: Human locomotion. | Human beings--Attitude and
    movement. | Posture.
Classification: LCC QP303 .L86 2020 (print) | LCC QP303 (ebook) |
    DDC 612.7/6--dc23
LC record available at https://lccn.loc.gov/2019025126
LC ebook record available at https://lccn.loc.gov/2019025127

Contents

Foreword

This book should come with a warning! It will challenge and eventually change the way you see and understand the body. But what else would we expect from a philosophizing world-champion wind-surfer and a fluid-moving aesthete? We can be grateful that they have teamed up together to provide us with a new perspective.

Martin, the philosopher, and Linus, the aesthete, point out that when changing views there is no wrong perspective—"being different is not the same as being wrong"—and this book certainly gives a new and refreshing perspective. The change in perspective and vocabulary means it might take the reader a few passes through the book to really hear, see, and integrate the significance of what they say inside. That time will be well spent.

I first met Martin and Linus in separate bodywork classes. They were both quietly confident deep thinkers, happy to sit back and observe until they had the full picture. When either of them did eventually ask a question or make a comment the class listened. Both Linus and Martin demonstrated the ability to take in new information, process it and re-present it with enhanced clarity and even greater depth.

That same dynamic is at play in this text. Linus and Martin have taken their experience and understanding of many disciplines and distilled it into "Movement Integration." The understanding of their principles will, when applied, free you from protocol-based thinking. They will take you from posture to movement and from anatomy to "ensomatosy"—their own word and gift to the world.

The philosophy within the book guides us toward an understanding of integration, a concept many have tried to define and which is captured beautifully within these pages through words, pictures and photographs. Integration is about embodiment, and is both inclusive and incorporative in the truest senses of each of those words: integration is about relationships and communication, freedom and expression. As your guides on the journey toward integration, Linus and Martin have provided language and visual tools that facilitate any therapist's exploration of what embodiment really means. Movement and breath tools are also explored both by the writers themselves and by a collection of contributors that you'll find in part III.

If you have traveled widely in your reading you will be familiar with many of the names on the list of contributors. Each has something different to say about their experience with movement and each, like the main authors, enjoys testing orthodoxy. This book is about challenging current models and is full of fresh ideas from fresh thinkers. Thereby the book achieves its aspiration because, in Linus's words, the "meaning of life is to be able to move freely in body and mind and to find someone to love."

Linus and Martin have moved boundaries that constrain us in body, mind, and soul. As an expression of their commitment to communication they have created *Movement Integration*. This innovative book will inspire you to move your body, it will expand your mind, and, trust me, you will love it!

James Earls
Author of *Born to Walk: Myofascial Efficiency and the Body in Movement*
London, UK
July 2019

Acknowledgments

Linus Johansson

The best way to organize one's thoughts, ideas, and concepts is to write a book. Finding the right words, creating sentences, paragraphs, and chapters, and illustrating it all is something that is very challenging and demanding and at the same time very satisfying and rewarding.

And every time you align your thoughts and embody your ideas in something concrete like a book, you go into a state of development. This book has been a very creative endeavor and together we have created two new concepts that we present here, both of which we are very proud. The first is both a concept and a completely new word, Ensomatosy. The second is the Color Illustration Model, which is a completely new way to explain and illustrate movement between structures.

When writing a book you may also find that you are on a mission. A mission to declare your attitudes and share your beliefs. I have come to realize that in this book we have given very few, if any, answers, for this is a book of principles, not of methods. What we have done is something far greater than just provide a few answers, which are always at risk of being rejected. Instead we have presented a perspective from which many solutions can be derived. Instead of explaining a definite absolute we present an algorithm that is highly sensitive to whatever you put into it and that will deliver solutions suitable to the situation in which you find yourself at that moment.

For me, this book has been yet another milestone in my career and I could not have done it without the inspiration and knowledge gained from my colleagues and good friends.

It has been an absolute honor and a great pleasure to work with my dear friend and colleague Martin. Martin has been, and always will be, my teacher and mentor in this line of work. To be able to be a part of his dedication to deepen and broaden the interpretation of the human form and function is a great privilege. For this I will always be thankful.

To my teacher Don Thompson, you were the first to truly introduce me to the holistic approach to the human form. Your dedication and devotion inspired me. I carry your words with me every day and they still guide me in my daily work and in the teaching I do. For this too I will always be thankful.

To my colleague and dear friend Cecilia, SOMA MOVE® has opened so many doors for me. It has tied loose strings together and during each session I always learn something new. I am very grateful to be able to do this with you.

To my mentor Gary Carter. The energy you give and the inspiration always experienced when in your presence is incomparable. I praise the job you do in enlightening the movement community and changing the scene completely. I am very grateful to have made your acquaintance and for all the things you have taught me, and will teach me.

To my teacher Julian Baker. I am deeply grateful for what you have shown me and given me the opportunity to experience. Your incisive questions have formed and shaped me and made me make critical changes in my work. For this also I will always be thankful.

We both want to give a special thanks to our teacher, colleague and friend James Earls who has written our foreword and given us much valuable input and feedback in creating this book. Please read James's book *Born To Walk* to get an even wider view of human form and function.

I also want to thank all the clients and patients that I have met throughout the years. You put your faith in me and gave me the opportunity to help and thereby let me clarify my principles and develop my methods.

I dedicate this book, and all my work, to my beautiful children and my beloved wife. I love you.

Linus Johansson
Linus is a physiotherapist, movement practitioner, and movement integrator. Linus works with patients in his clinic, presents workshops, and educates other therapists and trainers. Linus is the author of several books and together with Cecilia Gustafsson has developed the movement concept SOMA MOVE˚.

Martin Lundgren

I want to thank all the great teachers that I have met throughout the years. Special thanks to Thomas Myers, James Earls, Lauren Christman, Larry Phipps, Fiona Palmer, Kirsten Schumaker, Gary Ward, Julian Baker, Gary Carter, Jerry Hesch, and Jun Po Denis Kelly. I am also immensely grateful for the presence of Lena Björnsdotter and my dear colleague Linus Johansson. Thanks to all students for your interest and dedication. Above all I would like to thank all my clients—they

and their bodies are the greatest teachers and the true masters.

Martin Lundgren
Martin is a Board Certified Structural Integrator (ATSI/KMI) and movement practitioner. With roots in structural integration and Anatomy in Motion, he has been developing new treatment systems and treatment methods in recent years. As well as educating other therapists and trainers, he develops people's kinesthetic ability, performance, and well-being in his clinic through his way of working with treatments and movement.

PART I

Introduction

This book has been written with the aim of giving you an insight into, and understanding of, the perspective that we use to interpret and appreciate human form and function. It represents a milestone on our own journey of constant transition and development, and we acknowledge that our understanding today will most certainly evolve and develop in the years to come.

In this book we share what we—at this point in time—have found to be the most efficient and interesting way to explore and develop human movement potential. We do not proclaim this to be the absolute truth, but merely an interpretation of the vast complexity of human form and function.

This is a book of principles, and not a book of methods. This ultimately means that you will never read the sentence "If you see this, then do that" in this book. Instead we have created a new platform for you to be able to set your mind in a different direction or, if we

are lucky, to give you even more thrust and support in the direction you have already chosen.

We admit that we challenge the old paradigm in this book and for a good reason. We know that challenging what one holds to be true is one of the major measures needed to create progress and development. If we look back at history this has been proven over and over again. The fact is that if we are content to settle in and do not choose to constantly move forward, we will eventually start to go backward.

In order to be able to move forward, however, we will need to leave some things behind so as to create room for new ideas, thoughts, and progress. We therefore also invite you, the readers of this book, to challenge our principles and explanations, so that this, in turn, can lead to even higher development, for the benefit of all.

Please enjoy this book with an open mind and with a wish to participate in new ideas and insights.

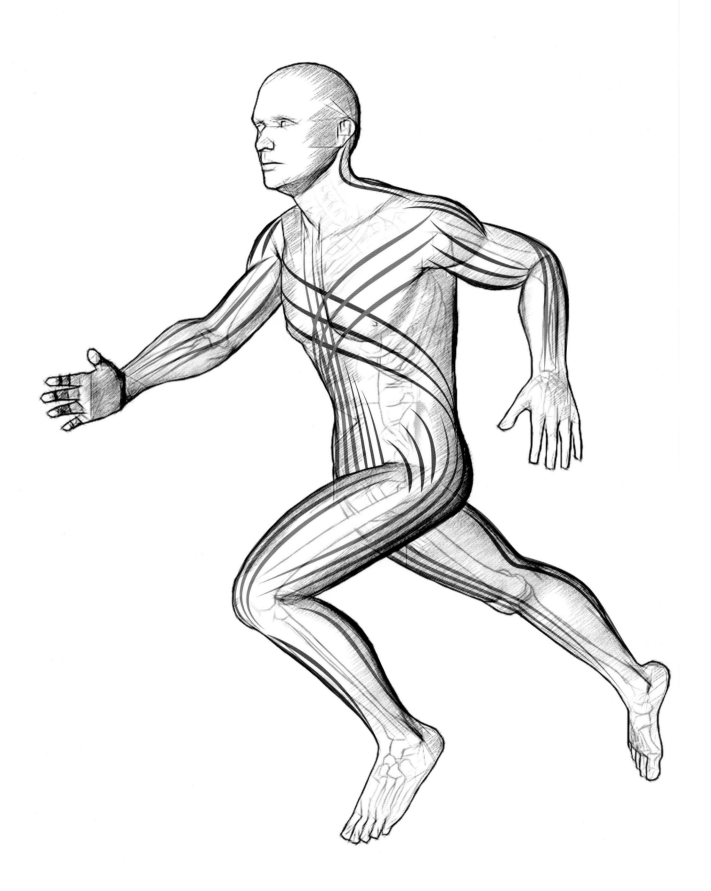

CHAPTER 1

The law of perspectives

Linus Johansson

*"Hold your good practices dearly and
your theories very lightly."*
—Gil Hedley

It is absolutely true that the human body is much too complex ever to be fully understood and described in a unifying way. Centuries of research and science in combination with the vast spectrum of therapy forms that exist today prove it. For if we knew everything there was to know research would stop and we would all unite around one form of treatment. At the moment this is not the case, and it probably never will be.

This is not a bad thing, though; it is actually something rather fantastic. Instead of one absolute truth—a monoculture, so to speak—this variety instead gives us a colorful and interesting spectrum of thousands of truths, a true polyculture. And if there is something we have learned about the human body and mind it is that it thrives in variety. The higher the variety of stimulus and input the higher the development of one's body and mind.

You know that the more varied movements and challenges you do the more you develop your body's potential. It is the same with your mind, in that the more you experience, the more interesting and varied the books you read and the people you meet, the more your mind grows.

This idea is something that we advocate in this book and we want to point out that our intention is not to present a new absolute truth. Our intention is instead to present one more perspective, another interpretation of the truth. With this book and its perspective we want to contribute and give another understanding of

the complex human body in structure and function. This will be one more angle of approach to help widen the possibilities available for to working with and developing another person's movement potential.

The principles of perspectives are important. The human body is endlessly complex and the only way to try and understand this complexity is to interpret what one sees and perceives through a chosen perspective. A perspective that gives an entry point toward this vast complexity and creates a platform to be able to address and discuss processes and developments that happen in the body when you work with it.

To be a true perspective it must be based on sensible principles, documented facts and sane reasoning. It also needs to prove to have good outcomes when applied practically. If it has all that, we are the first to acknowledge it.

Let us give some examples. One classic perspective would be that the body is composed of bits and pieces, such as muscles, bones and tendons, all with different names. The muscles are connected to the nerves that connect to the brain and they control the bones using the muscles connected via the tendons. This creates contractive movement in relation to the horizontal and vertical line. The skin, fat, and connective tissues are of no great interest and have little or nothing to do with movement. This is a perspective taught in most classic institutions evolving around the teaching of the body.

Another perspective would be that the human body is one indivisible and flowing unit, not built up from bits and pieces but grown from a single seed. No tissue or aspect in the body is more or less important than another and they are all connected and related to each other. Together they contribute to the loading of eccentric movement through gravity that is the foundation to all movement.

These are two rather different perspectives and set next to each other we can clearly see their differences. What is important to address here is that they are both correct. How can that be? How can two such different perspectives both be correct? This is for a very simple reason, which is that they are both based on sensible principles, well documented facts, and rational reasoning and, most important of all, they are trying to fulfill the same purpose, namely to give an explanation for the unexplainable.

However, it is easy to stand by one perspective and be true to that belief while looking at another perspective and seeing the differences between them, and therefore claim the other perspective to be wrong. One reason for this is that perspectives do not necessarily translate very well. They can use different terminologies and have completely opposing views on the same subjects.

When trying to communicate from one perspective to another the almost inevitable misunderstandings can leave the two parties quite confused and with the desire for each to claim the other to be wrong, based on the simple fact that they did not understand what the other was communicating.

"Being different is not a definition of being wrong."

Being different is not the same as being wrong. We must understand that a perspective is not the absolute truth, it is an interpretation. There are thousands of perspectives and each one of them has been constructed by a limited human mind trying to grasp the ungraspable.

One could claim perspectives to be the equivalent of religions. Each world religion is an interpretation of the world around us, constructed to make life easier by answering the big questions we have and to encourage

love, understanding, and affinity. This is fundamentally what all world religions preach.

In the same way, as long as a movement perspective is based on sensible principles, documented facts, and sound reasoning, with the intention of developing a person to their benefit, it can only be right, never wrong.

To underline this and to give this book a deeper foundation, we have invited a spectrum of different people to contribute to this book with their own unique perspectives.

Our intention is to let you see that one question, taken through different perspectives, will give a beautiful variation of answers and by linking these variations in our discussions throughout this book we want to give you a greater understanding of the polyculture we all live in.

To conclude, if you meet someone who clearly sees things from a different perspective than you, do not push their views away and claim their perspective to be wrong just because it does not go hand in hand with what you believe, or because you do not understand it. Instead, listen to their reasoning and try to see it from their perspective. Perhaps this person can present interesting theories and explain principles that lead to interesting methods. You might thereby actually broaden your own perspective and learn something new.

Foremost you will learn to understand that there are many more perspectives and interpretations of the truth than you perhaps knew before. To see and accept others' beliefs is what will make your mind grow and even start you questioning yourself and what you believe to be true.

We therefore end this chapter with the wise words of Gil Hedley that began it. He underlines it perfectly when he says,

"Hold your good practices dearly and your theories very lightly."

CHAPTER 2

The purpose of life and pain

Linus Johansson

For as long as humans have existed and evolved in to self-aware creatures, the question of why we are here has no doubt always been asked. You have probably asked it yourself. What is the meaning of life? Why am I here? What is the purpose? This is a philosophical question with as many answers as there are people, and every person has the right to state and believe in their own purpose in life.

But, what if we let go of philosophy and look at it all with more pragmatic eyes? What if we sharpen the question and ask: What is our true biological purpose in this life, according to evolution? Why has evolution taken us to where we are now and how is this now affecting us in this modern society?

Why is this question important? By elaborating on these questions, we can get the first insight to the understanding of the human body in structure and function.

If we make the assumption that evolution is based on DNA altering and driving the organism to be more optimized to functioning and thriving in its environment, two basic components are needed. The first is for one organism to be able to cross DNA with another organism of the same kind, or to evolve within the organism itself. The second is to be able to move. Movement is essential to be able to spread over an area to populate or to engage with another organism to reproduce. This can be a physical movement of the entire organism or just the movement of the spores, seeds, or roots. However you look at it, movement of some kind is required.

All this is true for the human organism too. We need to move to find food, shelter, and someone to cross DNA with in order to make our species survive. If we ask what the meaning of life is for all life on earth, according to evolution, the short answer would be: to move and to reproduce. Slightly rewritten for the human organism it would be: the meaning of life is to walk this earth for a while and to have children. Nothing more, nothing less.

This is the basic meaning of life and the fundamental and evolutionary purpose of your existence today.

From this basic definition we can take the discussion further and in a pure philosophical way elaborate on what our purpose is here on earth and what the meaning of life is. It is worth considering also that, as far as we know, we are the only organism on this earth that can view itself in this objective manner and ask these questions. All other organisms live under the principles that evolution involves without asking questions.

If we assume that this is the answer according to evolution, we can all agree that it is a very stereotyped definition and not in tune with the intellectual development that we humans have in this day and age. What if we were to rephrase this to achieve a more enlightened definition? It would be something like, "The meaning of life is to be able to move freely in body and mind and to find someone to love".

I am aware that a big serving of philosophy has been added into this new definition, but it is important to work with a concept that includes every single person in this world and their rights to think and do what they want and to love whom they want.

Having stated this enlightened definition we can now strip it down to one of its core aspects, which is the notion of *moving freely*. This core aspect of the definition is what we take off from and fall back on in this book. This is also one of the core aspects of the new perspective.

The simple definition that evolution gives us is just to move, nothing extra. Moving well or moving without pain are not of interest to evolution. Why is that? How come evolution does not have an answer to pain or help us to always be pain free? This is a question that is both interesting and important and if you doubt this, just take a look at yourself. Through millions of years, evolution has taken the human race to what we all are today. Highly advanced, self-aware, thinking, creative beings. And yet still we are tormented by the presence of pain in our lives. We have all experienced pain, the question is why?

What is pain?

No one truly knows what pain is, how it originates or where it actually resides in the body. Of course many theories have been presented and a great deal of research has been carried out to explain the phenomenon of pain. There are theories, but no one can say for sure. Pain can also come in many forms and shapes, from pure bodily pain to emotional pain.

All we know is that almost no one wants to be in pain. Still, over time, pain is present in almost everybody's lives in one way or another, both in movement and in other aspects. This is why pain is truly a baffling aspect of the human body. Few people in pain enjoy the feeling.

The common conception is that pain is a bad thing. But what if it is actually the opposite? What if pain is not something "evil," as we perceive it? What if pain is one of evolution's primary drivers for our unique existence and our ability to be the moving creature that we are?

To understand this we need to know what pain is to evolution. Like everything in your body you are driven by inputs and stimulus from your sensory organs. They are compiled together with your will to move and this results in the actions that make up your life.

Depending on the stimulus, you will react differently in different situations. If you see something you like you will react to it with positive feelings. If you hear your name called out you will turn your head. If you feel the warm touch of a loved one you will be relaxed. If you walk into a room and smell something you recognize from when you were younger you will instantly be taken back to a different time and place. All of these are examples of stimuli and how we react and adapt to them.

Pain is the same thing. It is an input, a stimulus that we must react to. It is a sensory input of touch, heat, mechanical alteration in structures, and when it passes a certain threshold it turns into the perceived sensation of pain. Being in pain or anticipating pain are therefore two great inputs that will alter how we interact with the world and how we interact with movement.

Let's elaborate. From experience we know that our direct actions can lead to pain. Walking on sharp objects, bumping our heads, or tripping and falling—all of these hurt. By anticipating what is going to happen, we orientate our movement to avoid being hurt. Imprinted in our genes is the subconscious knowledge that the action that will hurt us also can lead to injury and when injured we can no longer "move freely."

This is important to evolution and injury must always be avoided at all costs. Being injured and unable to move will render the individual unable to seek safety, or to find shelter, food, or a partner. That is why pain is such an important input and why we experience it as being so powerful. It is of utmost importance not to get hurt and injured if you want to survive, especially in the days when there was no support from medical care, when a sprained ankle or an open wound could lead to death. Today they are less serious, but still the pain is as present and powerful as it was back then.

By anticipating our actions we can try to avoid pain, for example by not stepping on sharp objects or hitting our head on the doorframe. These are obvious actions to avoid; however, not all actions can be anticipated. Examples might include tripping and spraining your ankle, picking something up and throwing your back out, or doing a high repetition movement with your arm and getting a strain.

You can also get hurt from not doing movements. A wrong position at work, such as holding tools awkwardly or sitting in a bad position in front of a computer screen, can also lead to pain.

Anticipating pain to avoid getting hurt in the first place is one thing, but when you are already in pain and certain movements you do generate more or greater pain, you anticipate these movements, both consciously and subconsciously, in order to avoid them. Your body will do anything not to stress the injury more with the risk of making the pain worse, rendering your body unfit to move.

The solution that your body uses is to "dodge" the pain by creating a compensatory movement. This means that the body needs to alter the movement that was expected and create a new movement pattern to avoid the pain, in a classic compensation.

The act of compensating is the solution that will keep us moving. This was crucial way back for surviving the day, but today it is just a hindrance and an obstacle in people's lives that therapists all over the world work with everyday.

We can contrast the situation when a car breaks down. It may stop and not move another inch because one single component does not function. When you sprain an ankle, however, your whole movement apparatus does not just stop because one part is "broken." Instead you create a compensatory pattern around the injury—with a sprained ankle you limp. Not the optimal movement but still movement, good enough to keep you moving forward. This was one of the primary solutions for our ancestors to keep moving and staying alive. In pain perhaps, but rather that and still be able move forward.

The key to compensatory patterns from pain is that they are based on the fact that all ordinary everyday movements, and by and large all movements in sports,

are executed based on the experience of doing the same movement thousands of times before. We don't need to think when we do them, we just wish for them to happen and they do.

For example, at the breakfast table, when you want to reach for your coffee mug you do not have to think that you need to abduct your arm, extend your elbow, and open each finger and each joint in your hand to execute the task. All you think is that you are in great need of coffee and all of a sudden you've grabbed the coffee mug. It all happened subconsciously and you did not have to think of every little movement in your body that was associated with grabbing the mug and moving it to your mouth. You do not even have to see it happen. You can be reading the paper at the same time and not letting your eyes leave the page and still manage to drink your coffee. This goes for all movements, including sports.

What happens when pain is present is that the body anticipates the pain when you wish for a specific movement to happen and quickly alters the movement pattern, from the old habitual pattern to a new compensated movement. Anything to avoid the pain.

Back to the example at the breakfast table: if you had some issues with your shoulder and were reaching for the coffee mug, your body would change the movement pattern and decrease or alter the movement over the painful structure in the shoulder and create a different movement elsewhere, in the associated areas, to let you feel less pain when moving while still completing the task.

This is an undoubtedly an acceptable solution for a short while. However, there is a drawback with compensatory movement patterns that are used over an extended period of time.

What happens when movement is altered from normal to compensatory is that the body uses other structures, or uses the ordinary structures in a different way, to perform the task. The problem quickly arises because these structures are most likely not optimized for the task or the new version of the task.

This can lead to some major issues. One is that the other structures used to perform the task get overloaded and fatigue if the compensatory pattern is used over an extended period of time. This, in turn, can lead to damaging the structures, creating new pain and even injuries.

Another issue is that the body can alter the range of motion over a structure as a solution to protect it and make it less painful. This alteration is often a decreased range of motion over the painful area. In the long run, these compensations can lead to a couple of common outcomes.

As the structures are closely dependent on each other, to make the body one functional unit, the lost range of motion in one area can inhibit the person due to an overall sensation of accumulated loss of mobility in the body. This can in turn lead to kinesiophobia—the fear of moving—and can altogether render the person more sedentary and avoiding physical activity. This will then lead to even more loss of range of motion in other structures due to the lack of physical activity, starting a bad trend and leading to more pain and lifestyle-related diseases.

On the other hand, if a person with a loss of range of motion over one specific structure is still active and moving, the body will react to this. The reaction, unfortunately, is not regaining the range of motion in the lost area but instead increasing the range of motion over another structure that more easily yields for the forces transmitting differently in the body due to the compensation.

The following is a classic example that we, as polyclinic therapists, face all the time. A person injures a structure in the body during a specific movement—let's say the foot whilst running. The person is in pain and rests from running for a while and limps through his ordinary life, working, interacting in light activity and playing with the kids. Over time the injury leads to a loss of range of motion in the affected area, due to a complex combination of structural damage, pain, and compensatory movement.

Time goes by and frustration, in combination with the body's ability to lessen the pain in the foot by compensatory movement, soon gives the person the idea of picking up running again. Unfortunately, all too often, the person's mindset is at the same place as before the accident happened and has not adapted to the fact that the body is somewhere else in structure, function, and capacity.

The person sets off with his plan to start running as before, with the preconceived idea that the body should behave just as before. The difference now is that the capability of the structures has changed for the worse, because the person has been inactive.

All too often, the injured person does not undertake any rehab or treatment or take time to investigate the underlying reason as to what caused the injury in the first place. There is often no interest in getting involved in their own body. We are all so used to having a body that just works and do not think it can be any other way.

The person soon realizes that the pain has not completely gone away in the foot, but still this person keeps struggling, workout after workout. Soon a new pain arises in the body, now in the knee on the same side and shortly thereafter that old back pain from five years ago is starting to resurface. All this leaves the person in despair over what is happening and they now seek help. This is an all-too-common scenario and a classic "not listening to my body" story.

What lead the person to not listening to the body was perhaps that there was a big race coming up and they needed to get enough miles in the legs before being able to participate, or perhaps it was that classic anxiety-driven desire to lose weight that pushed them to continue to run even when in pain.

As the body wants to keep performance and function as constant as possible it will, when locally lost, delegate mobility and function to other structures in the body via compensatory patterns. Thus, losing the mobility and function in the foot and still choosing to run will set of a wave of compensatory patterns up the movement chain. In this classic example, mobility will be moved to places such as the knee and back to mimic the total expected outcome of movement in the body when running, with the risk of creating more pain down the line.

Back in the day, our forefathers did not have a race day coming up nor were they plagued by the anxiety to lose a few pounds around the waist; it was the sheer will to survive the day that drove them to keep moving on. If they stopped they fell behind; pain or no pain, survival is the driving force of evolution.

The intriguing thing in all this is that at some point in human development we became self-aware and could give expression to pain. We could experience pain on a conscious level and at that moment we started to try and explain and give words to the experience. We can see that we have also tried to find answers and solutions to the phenomena of pain. It was probably at that same point in time when pain started to play mind games with us, and it has done so ever since.

In its pure and evolutionary form, pain was a key to our survival, one of many sensory inputs that was part of our ability to survive and to procreate. Not unique in any way to humans, all life forms that can perceive pain use it as one of the major sensory inputs to stay alive.

Today, pain is still a mystery. It is a mystery for the person who is in pain and for the therapists, doctors, and scientists that try to understand and treat pain.

Stepping away from the past and into our modern lifestyle we can see that a new kind of pain has emerged, a pain that probably our forefathers never experienced. The price of self-awareness in a fast and highly demanding modern lifestyle can manifest in another form of pain in the body, namely psychosomatic pain. This kind of pain is perhaps the most complex pain of all. It arises from trauma to the mind and manifests in the body as pain. It is one of the most delicate forms of pain to treat and it is unfortunately all too often neglected for that very reason.

It is neglected by the patient because it is too difficult to face and manage one's entire life situation rather than to do a couple of exercises for the neck and shoulders. It is neglected by the therapist as it is "mission impossible" to fix the devastating headache that comes each time the person experiences stress with a few exercises.

> *"Because pain changes everything."*
> —Gray Cook

Nonetheless, pain is pain, be it from tripping and spraining an ankle, a herniated disc, or a complex psychosomatic pain. As Gray Cook puts it, "Because pain changes everything," referring to the movement patterns, and, however the pain resides in the body, we will always try to dodge it by altering how we move.

Bearing all this in mind, what is truly daunting is the fact that you are alone with your pain. No one can ever feel or experience another person's pain. All pain, in all its shapes and expression, be it in the body, mind, or soul, is an individual suffering and we all deal with pain in our own way.

This is why another person's pain, always, must be treated with the utmost respect and never be diminished by someone else. To diminish another person's pain is to diminish them as a person altogether.

We can conclude that what we describe and discuss here is based on much more complex functions and relationships. We use explanatory models to achieve a simplified picture of these vast and maze-like connections that lie within each person's body.

Everything happening in the body, be it pain or movement, is based on an intimate communication between the body and the mind. A communication that no one possibly will ever be able to truly understand. The reasons why we never will understand are many and one of them is the paradox of trying to look objectively at a subject and at the same time be the subject. We are like a puppy chasing its own tail. Doomed to run in circles forever.

This leaves us with one solution, if we cannot find exact answers we have to make an interpretation. This brings us back to the beginning of this chapter and the enlightened definition that we made of the purpose of life and the core aspect that we extracted, "To move freely."

Therefore we will continue this book, orientating ourselves around the idea that we *should* be able to move freely. Free from pain, free from compensations, and free from mental or systemic obstacles. Free to move effortless through gravity and will. Even if evolution thinks otherwise.

CHAPTER 3

One foot in front of the other

Linus Johansson

"To understand where we are today is the only way to know where we are going tomorrow."

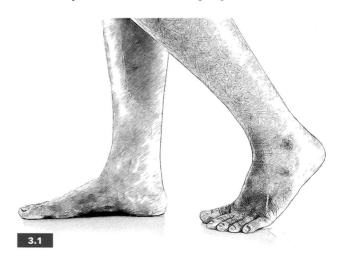

3.1

When did you ever look at yourself in the mirror and realize that you, as a mammal, have a truly unique "design"? This design creates some very special properties, especially in how you move, for you are the only mammal to ramble this earth on two legs. Sure, primates can walk on two feet but they would rather not, and they definitely cannot travel as far as humans when doing so.

Yet another aspect that is unique to you and your design is something you do not see when you look in the mirror. It is an underlying quality that gives you, as a bipedal creature, a very *special* quality: you have a body that is highly energy efficient. This means that when you walk on your two feet, you spend very little energy in relation to the distance you travel.

You actually have experienced this quality within you many times before, but probably not thought much

about it. If you recall that beautiful spring day when you decided to take a walk in the lovely weather. You walked mile after mile without stopping to rest. When you got back home you didn't feel exhausted, instead you felt invigorated. You seemed to have more energy when you got back than when you set off. It is an interesting feeling and an interesting fact that when you move as evolution intended you can accomplish great distances and feel refreshed doing so. Of course you spend energy taking a walk and the feeling of having more energy is just a subjective experience.

What is even more interesting is that the more often you move, walk, or run the more energy efficient you become and the better you feel. You adapt, learn, and develop your body and its efficiency, and you want to do it more. What is intriguing is that this also is a big part of the human evolution and existence. Only one of many aspects, becoming extremely energy efficient is probably one of the major features that made us what we are today.

The human body's rhythm of movement and interaction with the ground beneath through gravity is truly something very special. Trying to grasp and understand this is a true commitment. That said if one does commit, in the light of this new perspective, it could truly be rewarding. And the more one learns to appreciate the fact that we have been shaped to stand upright on our two legs and float with ease through gravity, the more we can start to develop our own and others' capabilities and abilities.

To start this journey we need to understand how this all happened. How did we evolve from walking on all fours to standing upright? What part of evolution invited us to become bipedal?

The theories on how we evolved to become a walking creature are many. Daniel E. Lieberman presents a very plausible theory on how we became walkers and runners in his book *The Story of the Human Body*. The idea that we freed our hands to stand on our two legs to reach fruit from higher branches and make tools is a long recognized theory. But is it all that plausible?

Daniel presents an alternative idea in his book. It was probably a strong outside stimulus, as it usually is, that triggered this phase of evolution and was the start of the journey to where we are today. That fact that we did free our hands from the ground and became able to make tools and pick fruits from higher branches probably happened parallel to or after the development of us standing erect, and for a very different reason.

The theory that Daniel presents in his book is that our ancestors experienced a long period during which the climate changed. This led to feeding areas being more and more scattered and to food getting harder to find. Our ancestors had to start moving over greater areas than before in order to find something to eat. This in turn led to the demand for our bodies to become more energy-efficient when moving, since our ancestors would not survive if they used up more energy moving to a new feeding area than they could acquire from the food they obtained there.

This led to our forefathers evolving a pelvis that made them, over time, move more upright and eventually get up from all fours and become bipedal. This allowed them to move over greater distances and use less energy to do so. With the evolutionary step of becoming bipedal and acquiring a more upright position came also a lot of new and specific movements.

One of those movements was being able to, in a more refined way, center over one foot when moving forward on two legs. This was, and still is, a unique feature for us humans. Primates do not have the same construction of the pelvis, as we have, and therefore need to shift their entire upper body from left to right to keep their balance when they walk on their back legs. This is not an efficient way of moving, and according to Daniel,

a chimpanzees only walk 2–3 km per day as they spend four times as much energy when moving than we humans do.

3.2

We humans have the ability to find our center of gravity over one foot by shifting our major structures in our bodies in relationship to each other and not as one unit over the standing foot. The benefit with this maneuver is that it does not throw us off balance and does not require big inefficient movements. This feature, in combination with well placed myofascial tissue that can store kinetic energy when moving through gravity, makes us the "optimal movers" we are today. Remember, we have populated almost this entire planet by walking from one corner to the other. That's one great stroll.

Regardless, whether we believe that it was the development of the opposable thumb or climate change that drove us to evolve and stand up on our two legs does not really matter. We are where we are today, however we got here,

3.3

and we cannot really do anything about it, but just appreciate it.

From an individual point of view, both in personal and functional development, we can probably agree that knowing where we came from can be interesting—but not half as interesting as knowing where we want to go and understanding what we have to do to get there. It's like reading a map—if you don't know where you are you can't get where you want to be. In this book "knowing where we want to go" implies the journey of development of the physical body in movement and the intellectual mind in awareness. And to accomplish this we need principles that we can use to inform our choice of methods and arguments, and fall back on when we seek answers and guidance in the challenges that will arise.

That is why looking back to understand where we came from, and at ourselves now in the present time, gives us the understanding that we are structurally evolved to do one thing, which is put one foot in front of the other and move forward through gravity.

Understanding that this, the gait, is fundamental to our very being gives us the foundation of our principles and thereby the means to build a high and strong structure for this perspective of the human body in form and function. (Please read Chapter 23 by Gary Ward for a closer look and for his understanding of the gait.)

One might ask, are walking and running really the only major movements we are meant to do? The answer would be: Yeah, kind of. Especially if we consider evolution: but how can this be? We do all these other amazing movements as well: squatting, lunging, deadlifting, sitting, rolling, swimming, hugging, kissing, making love. Yes we do, but the key point here is that we don't have to. All of these are actions you choose to do because you can, not because you have to. You can live a whole lifetime without doing a single push-up, handstand, or deadlift and even without kissing a single person. Thus, you can choose to do certain movements in life and avoid others, but one thing you can never choose not to do is to put one foot in front of the other and repeat. (If you are born well and healthy that is.)

From the moment you were born your body started to respond to gravity. From that weightless state in your mother's tummy to suddenly being thrown into a force that kept pulling you down, or pushing you up, depending on how you interpreted the force of gravity. All through your first years your evolutionary heritage stimulated you to explore gravity and to load your system—roll off your back, get up on all fours, stand tall on two legs, walk then run—and you did it because your forefathers did it and passed it along down the evolutionary line.

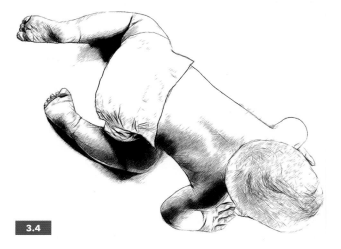

3.4

You followed that gene code that is deeply engraved in you through generations and stood up on your two feet and walked. No one taught you this, you taught yourself to move as evolution intended, as do all life forms on this planet. They follow their evolutionary heritage and you are no exception.

Why is this important? If we are meant to walk but we can do all these other movements, why does it matter? Because the gait is the foundation of human movement, and all other movements that you can choose to do are based on the rhythm and construction of the gait. This is interesting especially if we go to the "churches of movement," which at the moment are fitness and workout establishments.

The old school tells us that each movement has its own properties and that we need to address them accordingly to be able to perform the movement correctly. Thus, if the knee hurts when I run, if my back aches from lifting a barbell, or if my squat isn't deep enough I need to work on those movements specifically, one by one, and the different characteristics for each movement.

The conclusion is that the old school let us believe that each movement is unique and that different movements—e.g. squat, run, jump, swim, throw—don't have anything to do with each other, and especially not with the gait.

Segregating movement this way is simply an expression of trying to understand and simplify the complexity happening when we move. This is because we humans have always been searching for answers regarding all kinds of topics, whether for ourselves or the world around us. It must be said that trying to understand can never be wrong. Describing what we see and experience around us is what drives intellectual development. Therefore we do not imply that what the old school teaches us is wrong. We acknowledge it as one of many interpretations of the truth and will never claim it to be wrong. We are also fully aware that our perspective is also an interpretation of the truth.

We believe that everyone is free to construe the human body, in structure and function, from any given perspective, as long it is a perspective based on sensible principles, documented facts, and sound reasoning, and with the intention of developing a person for the better.

There is no question that the human body can perform outstanding movements and feats. From our perspective, however, the question is whether they are really different movements? Or are they all the same movement just put together in different versions to become that specific and unique movement?

When we look at movement being performed we see different versions and compilations of the rhythm of the gait. Our ability to flex, extend, abduct and adduct, rotate, shift, tilt, and bend our structures are all there to help us perform the gait, to make us perfect, energy-efficient walkers and runners.

This can be both a bit daunting and hard to grasp at first. Bear with us and this will soon become more clear and logical to you. And if you choose to start looking at the body from this perspective we promise that you will see and understand things in a completely different way than you have done before.

You still might think that it is a bit far-fetched to claim all movements arise from the gait. What about the opposable thumb and the very high freedom of movement in our shoulders? How are they connected to the gait?

Our opposable thumbs and mobile shoulders are functions that we developed thanks to the fact that we became bipedal creatures. All four-legged creatures are dependent on their front legs, and had we too stayed on all fours we would probably never have developed these qualities in these structures. The hands, arms, and shoulders are very much bound to the rhythm of the gait as we once used them to move and still do. Have you ever walked without swinging your arms? Imagine the Olympic 100 m finals taking place with the runners' arms taped to their sides, so they are unable to swing them to create that rotational torque. Sure the runners would cross the finish line, but no world records would be broken.

In conclusion, if we can see and understand the structure of the gait, we can see and understand all other movements that the human body can perform. This would imply that if we are failing to perform one of those thousands of other movements that we can choose to do we can always go back and assess how this failure is manifested in the structure of the gait. We can then address that failing movement and reintegrate it into gravity.

The argument that "it's all connected" will prove to hold if we can create an intervention based on the structure of the gait and show that it will improve not only the failing movement in question, but also bring the entire body to a new state of total body function.

In truth, however we first got up on two legs, freed our hands, and became self-aware is anyone's guess. We respect all beliefs and theories contributing with an explanation. Nonetheless, we all walk this earth in the same way and gravity loads our structures in the same synergy when we move. Therefore, to truly understand the potential in human movement one must learn to see and appreciate the rhythm of human movement through gravity. To explore and understand more of our movement potential on two feet please read James Earl's book *Born to Walk*.

CHAPTER 4

The principles of a new paradigm

Martin Lundgren

Traditionally, the study of the human body, and in particular the study of human movement and structure, have come from a reductionist perspective. The implicit premise in this perspective is that it is possible to get a complete picture of the workings of the body by reducing complex phenomena into simpler parts and then studying the parts. In the western world, most science and medicine is based on this premise, and it is a big part of our heritage. Of course, it would be absurd to argue that this perspective and what it has generated do not have any value. However, what we do claim is that when it comes to our field, which can be called "the field of human movement and structure," this reductionist perspective can be severely limiting and sometimes serves as an invisible conceptual constraint, hindering progress and preventing us from developing things further. We therefore see the need for a description of another perspective, or another paradigm, that we feel better represents our conceptual framework and our methodologies when it comes to working with the body. That is what is presented in this chapter. To make things more understandable, we will compare the traditional reductionist paradigm with this new paradigm, which we will call the systemic paradigm.[1]

The trouble with language

It comes with hidden premises and assumptions

When we do anatomical dissections we cut up the body to make it more understandable; we create anatomy through dividing and cutting up into smaller pieces what used to be one single piece, a whole. And when we use language, we dissect reality—with our minds. To make things understandable, we make abstractions and divide the world through our concepts.

Our everyday use of language has mainly evolved from a pragmatic way of interacting with the world that does not necessarily reflect the true workings of the world. The difficulty and complexity of language can be quite noticeable when trying to learn a new field or subject, where more nuanced concepts and language have been developed to more accurately reflect the underlying reality. This refinement of concepts and language has been very prominent in the sciences and the western world, going back at least to the Enlightenment in Europe, or even further.[2] It has given tremendous value to us human beings. It comes hand in hand with logic and reason, giving us the ability to distinguish between fairytales and scientific verifiable theories and concepts, which we think better represent reality.

The firmer insistence on logic and reason also gives us a chance to make an epistemological distinction between the things we know we can know, and the things we know we cannot know. We can positively point out to somebody that they do not know what they think they know because their concepts do not make any sense and their conclusion is invalid because it does not follow their premises. When somebody says that it is raining today because a demon is doing the hula dance in the closet, we can dismiss it as fanciful because the concept of a demon in the closet does not appear to reflect reality in any way, and furthermore, there seems to be no connection between the demon doing the hula dance and the fact that it is raining.

So what is the trouble with our increasing capacity to implement the use of logic and reason where it is needed? What is the problem with refining and clarifying our concepts and our language? Is that not a good thing?

The trouble comes when we think that the concepts and definitions we have created are free from our assumptions and hidden premises. If we cannot see the assumptions in which we are embedded, this can create a lot of problems. Moreover, even if we are aware of the assumptions it is really easy to proclaim them as self-evident facts without further inspection. This neglect sets up the stage for ignorance and naive certainty, thinking that we have a more complete picture of reality than what is actually the case.

In the reductionist paradigm, the deepest metaphysical premise is that the world is a world of objects. Of course, this is self-evident—how could it be otherwise? The correlating premise is that the defining we do and the concepts we come up with directly represent an objective reality. Because these concepts directly represent reality, there is no room for the view that language comes with hidden assumptions and premises—the concepts are either correct or false. The defining is a direct inventory of the world and therefore does not allow any subjective element. This way of defining creates an either/or logic where contradicting statements cannot coexist. The reductionist paradigm has a hard time handling the complexity of the world, and the most significant mistake comes when people embedded in it dismiss things as fantasy, mistakenly pointing out that some concepts are wrong because they contradict other concepts that we know to be "true," and dismissing the conclusions as invalid because they do not follow the known way to deduct things through simple logic and analysis. They are failing to see that the concepts come from another set of assumptions and premises and that the reasoning supporting the conclusion follows a higher order of complexity and logic. Because of this mistake, a lot of things not congruent with the reductionistic paradigm are dismissed as nonsense, the preference being given to supporting the status quo and keeping the viewpoint intact. This dismissal also makes it easier to keep away any anomalies created by the complexity of the world as well as to maintain the illusion of absolute objectivity.

In the systemic paradigm, on the other hand, the premise that is held deepest is that the world is irreducibly complex; if anything the world is more a world of processes than a world of objects. From this point of view, it is, therefore, better to take a pragmatic stance, exploring and taking advantage of as many perspectives as possible. It involves trying to accomplish the impossible feat of synthesizing the different perspectives and trying to handle the resulting complexity armed only with the limited cognitive ability of homo sapiens, with the intention of getting as complete a picture of reality as possible. When exploring different perspectives the reductionist perspective might be one of them.

When we know the limitations of language and our minds we might be able to use language in a better way. When we understand the difficulty of constructing adequate concepts, and know that they are biased by the fact that we are humans, and that they come with hidden premises and assumptions, we might be able to hold a more realistic view of reality, one that is less rigid and more open to development and advancement.

In the systemic paradigm, there is an awareness that concepts are constructed and are more approximations of reality than a direct representation of reality. This paradigm is not a constructed relativistic view where concepts are seen as wholly independently constructed in our minds, even though we might find instances of that, but a more a realistic view where concepts are constructed but dependent and constrained by reality and our premises and coherence with the rest of our conceptual framework. In this regard concepts can reflect reality to a greater or lesser degree; some might be categorized as directly false, but the simple dichotomy between true and false does not exist in the same way as in the reductionist paradigm. The more interesting thing to know is what function the concept has, what role it fulfills, what is its relationship to other concepts, and if is it congruent and coherent with the context and our stated premises and assumptions. And of course, we also want to see to what degree the concept reflects reality. Some apparent contradictions might disappear when we manage to synthesize different perspectives and create concepts that reflect reality in better ways.

The concept's degree of objectivity, of course, varies between different fields; physics is perhaps the most objective while other fields, for example sociology and psychology, can be found at the other end of the scale, with less objectivity. Even though physics plays the game of complete objectivity, it has to obey some metaphysical claims and assumptions as its starting point. That physics has to comply like this might not be so apparent with the classical Newtonian physics but has become more evident with quantum mechanics.[3] It would be the reductionist's dream to be able to reduce every field to physics, or at least biology to biochemistry and biochemistry to physics. We would then have one dominating consistent perspective on reality and not be forced to take differing perspectives with a different set of assumptions and having to deal with the unsynthesized complexity of the world. From a systemic point of view, reality is one whole thing, but it is too complex to reduce it to one field, and attempting to do so might result in a considerable loss of value. That does not mean that there is no dependency or relationship between different fields. The only way to get a more complete picture of reality would, therefore, be to explore different perspectives with different premises and contexts. The complexity in the world forces us to take different perspectives and views hopefully enabling us to create some form of metaperspective and a better view of reality.

The study of the human body is a good example of this. It is the perfect crossing point between different fields. With the field that we feel we belong to, the study of human movement and structure, there is a contingency and relationship to every field, physics, biochemistry, biology, psychology, and even sociology. If there were to be very big changes in any of these fields, it would also affect "the field of human movement and structure."

The body is a system of systems within complex systems

In the reductionist paradigm, because the world is a world of objects, the body is more or less seen as an advanced machine. In this sense the body is nothing more than the sum of its parts. There is not really any place for emergent properties of the body that come from the body as a whole system. Instead, all the properties of the body as a whole can be deduced from its parts and the interactions of its parts. Usually, this is done in a linear additive way. DNA and genes are usually seen as the dominating causal agent, working in one causal way, from DNA to RNA to protein, which creates some form of function, and most things in the body can be reduced to this.[4] There is very little room for subjectivity in the reductionist paradigm, as the whole point is to describe reality from a perspective of complete objectivity. Everything is looked at from the outside, and even subjectivity is usually reduced to an external view of subjectivity.

In the systemic paradigm, the view of the body is very different. The body is a meeting point where different systems merge and are contained within one whole. This whole stands in constant relationship to other complex systems outside of itself; it is in this interface that subjectivity arises, and it is also in this relationship that it is relevant to talk about emergent properties of the whole. In the reductionist paradigm, it is common to objectify this whole by separating it from the external relationships that exist, treating it as an isolated object more than a responsive dynamic process that cannot be separated from its context. From the systemic point of view, it is impossible to separate this whole from its constant relationship with other complex systems outside of itself, and if the body were to be taken out of its context, it would not be the same body. Also, the relationship between the whole and external complex systems constrains and affects the systems within the body. This means that if you study a system in the body in isolation and from this draw conclusions for the whole, you are going to have a less complete view of the body, a more mechanistic, static, and deterministic view. If, on the other hand, we were to take more of a systemic perspective, acknowledging the system as a whole and its relation to other external complex systems, we would get more of a dynamic, responsive, adaptive and even intelligent view of the body.

If we come from a systemic point of view, if some in vitro studies of cells in a petri dish contradict how the living cell seems to work in a body, this would then be less of a surprise. In the same sense, if we come from a reductionist perspective, an insistence on treating the disease instead of the person would just be a logical consequence of the perspective from which we are operating.

Let us take epithelial cancer as an example, which is said to account for more than 80% of people dying of cancer worldwide. Systemic chemotherapy treating advanced epithelial cancer can often lead to "dramatic tumor responses." Unfortunately, this does not translate into a longer survival for patients.[5] Chemotherapy from a reductionist point of view makes more sense in this regard as it is effective at treating the disease (momentarily). It makes less sense from a systemic perspective as it seems less supportive to the system as a whole.[6]

Within the systemic paradigm, every level of organization has the same ontological status, that is, one level is not seen as more "real" than another. Furthermore, it is not only one level that has causal potency, but every level also has the same chance to affect something on its own level or affect, constrain, or relate to something else on another level.[7] Because of this, subjectivity is not reduced to something insignificant but instead elevated to something that can work as an organizing principle with agency and causal power. In the reductionist paradigm higher levels of organization and subjectivity are usually seen more as epiphenomena, or secondary effects, meaning they do not have any causal power. Having this viewpoint and paradigm as an unexamined premise as a basis for human existence does not really play in favor of our human potential, to say the least.

Different views on causality

One basic difference in the different paradigms is in their views on causality. There is a difference both in the complexity of causality and in the nature of causality. In the reductionist paradigm, there is usually less complexity when it comes to how we think about causality. Causality here is usually characterized by wanting to find the single, most significant cause that brings about something else. In the case of a disease, for example, there is an intent to find the "thing" that causes it. If, for example there is a disease that can be easily traced to someone's genetic make-up, there is one clear, isolated cause. A broken tibia caused by falling

downstairs also has one clear, isolated cause. To add complexity to the reductionist paradigm, we usually do that by just linearly adding causes. We can, for example, have three different causes that together create a disease.

One question, "what is THE cause of my ailment?", is one that is frequently asked in my clinic. The client expects me to produce an answer that aligns with their question. However, in the systemic paradigm, causality has another degree of complexity. It is not seen as one isolated "part" affecting another isolated "part"; instead, there is an emphasis on the dynamic and complex interplay both between the different parts that, rather than being isolated, stand in a constant interdependent relationship with each other, and also on the constraining and affecting properties of the whole system affecting the individual parts. Put in different words, rather than having one isolated cause, the interest lies in the dynamic interrelational interactions between different elements in a system, the constraining emergent properties of the system as a whole, and the interrelational aspects the system as a whole has to other complex systems in its environment.

The nature of causality looks very different in the different paradigms. Causality in the reductionist paradigm is usually seen more mechanistically, where isolated static parts affect other isolated static parts in some way to create an outcome. A crude metaphor for reductionist causality is one billiard ball hitting another—anything that is not in direct contact with this billiard ball will not be affected It is an isolated, mechanistic, linear view of causality. In the systemic paradigm, there is more of a relational causality seen from a systemic point of view. This means that each part has a relationship to all the other parts. That relationship might look different and might be stronger in some parts than others, but still, a change in one part is reflected throughout all parts, and single parts also share a relationship with the system as a whole. In the systemic paradigm, causality does not happen in isolation. If we want to find a suitable metaphor for systemic causality it would be some kind of network interaction where one action is reflected in the whole network; this action might be more interlinked with some of the parts in the system and less to other parts. When one action happens from a systemic point of view, the system as a whole is changed—it is not the same as it was before. Even if the change is very small, the system is not the same. This continuous process of change makes up the dynamic behavior of the system, and if there is a need to define the system, it is better to look at recurring patterns of behavior rather than trying to create a static picture of the system. This differing view on what constitutes change forms a considerable metaphysical divide between the paradigms, one where the reductionist paradigm has more of a static, mechanistic, linear view of the world and the systemic paradigm sees change as an intrinsic part of the world.

Given this difference in views of the nature of causality, it is not hard to see that the complexity when it comes to causality differs substantially in the different paradigms. Because the parts in the systemic paradigm are not isolated but stand in constant relationship to each other, the possible relational arrangements—the "possible possibilities"—are substantially higher in comparison to the complexity in the reductionist paradigm.

The difference in how we divide and define

The way we divide and define parts also differs between the two different paradigms. In the reductionist paradigm, as soon as we have defined a part, that part will always stay the way we defined it. The part is static, and the properties of that part stay more or less the same (if it does not break down somehow). The defining of the part can be seen as an ontological statement, i.e. the defining is not descriptive but is an actual statement of what and how that part exists in the world; as we said before, it is an inventory of the world. The definition is conclusive and excludes any other definition that is contradictory. Because the definition of a part is an ontological statement about the world, the definition is static, for if we change the definition there must also be a change in reality itself. This creates an either/or logic, which means that either a part is one way or the other. Either a part is round or it is square; it cannot be both, contradictory properties cannot coexist.

This way of defining and dividing works very well when it comes to explaining phenomena that are purely mechanistic in their nature. If, for example, we want to know how a clock works, it makes sense to dismantle the clock, name all the parts, and see how they all work; there is no ambiguity in how we divide and define the different parts. However, if we try to explain organic things such as a tree or a human body, the dividing and defining is not so straightforward. What we choose to define as a part can then seem a bit more arbitrary.

In a systemic paradigm, the way we define and divide things is dependent on the context. There is an understanding of the limits, constraints, and difficulties in defining something. In this paradigm, it does not make any sense defining something if we do not specify our premises, views, and perspectives. Pointing out which kind of perspective we hold is recognized as crucial, as this is seen as an integral part of defining something; a definition might only hold true under certain perspectives and conditions, and fall apart under others. It is essential in this paradigm that a definition is coherent with the underlying premises and perspectives. In this paradigm, one significant element of defining something includes the relational aspects, that is, the connections a part might have to other parts. The relational element in this way of defining makes the definition in itself less static. If the context and the relational aspects change, the properties and behavior of the part might change, so in some sense, the part is more adaptive, or even more "intelligent." Because the part is more adaptive or constrained by its relational aspects, we might get properties or behavior that is contradictory when the part exists under different conditions and relational arrangements.

Emergent properties in a system

As mentioned before, in the reductionist paradigm there is no room for emergent properties that come from the body as a whole system. That is not the case in the systemic paradigm, and we will now discuss these emergent properties from a systemic point of view. These properties emerge from the whole and cannot always be reduced to the parts of the whole. One of the reasons for this is that the properties emerge within the relationship the whole has to complex exterior systems. The emergent properties might be dependent on the parts, but the emergent properties cannot be reduced to the parts. Studying the parts in isolation, without looking at the system as a whole and what relationship it has to other external complex systems, is insufficient to conclude anything on these emergent properties. One typical example that is usually given to illustrate this is water (H_2O). Water consists of hydrogen and oxygen, but the study of oxygen and hydrogen in isolation can tell us nothing about the wetness of water.

In systems biology one emergent property usually mentioned is robustness. Robustness is a property of a system at a system level, meaning that it can only be understood if looked at from a systemic level. In a robust system, the system can uphold critical functions within the system despite various perturbations.[8] If we put it in the language we have used before: the system as a whole adapts to other changing complex external systems through an adequate responsive behavior. The behavior of the system might seem logical from the viewpoint of the system as a whole and its current relationship to other external complex systems. However, if we only look at the current state of the system by examining its parts, we might just conclude that there is something wrong with the system, or something wrong with some of the parts of the system. If we were to draw some of your blood and conclude that you have some parameters that are "off", and we better try to get those markers where they belong, we might fail to see that the markers are simply an indication of an adequate behavior of the whole in response to external circumstances, and we might just end up doing something unsupportive to the system as a whole. An example of this is a study from 2008 that concluded that trying to regulate blood sugar levels more intensively led to a higher risk of dying, "As compared with standard therapy, the use of intensive therapy to target normal glycated hemoglobin levels for 3.5 years increased mortality and did not significantly reduce major cardiovascular events."[9]

If we look a bit more closely at our own field, the study of human movement and structure, one of the more interesting relationships we have is the relationship between the body as a whole system and gravity. If there is one relationship we cannot escape it is the body's relationship with the earth itself (if we do not travel to outer space). This relationship is both a constraint and a necessity for the existence of the living body as a system. We are now going to look at emergent properties within this relationship.

When talking about the body with clients or with people in general or sometimes with colleagues, you can come up against concepts that usually are ill defined and would need a healthy dose of the aforementioned clarification and refinement. When somebody says that something is "strong" or "weak," or a muscle is "short" or "long," there is usually a need for making those concepts more nuanced and well defined (not to say that some people might use them in a well-defined manner). The people you talk to then sometimes get a bit surprised when you swop the concepts altogether and try to represent more of systemic perspective. From our point of view, it is usually more interesting to engage in a conversation with concepts that come more from a systemic perspective. One of the concepts that represent an emergent property that comes out of the relationship between the body as a whole system and gravity is the concept *integration*.

> *We define integration as when intrasystemic elements communicate and cooperate in such a manner that it allows for greater integrity and efficiency, which creates a higher order of complexity and functionality in a system as a whole in relation to gravity.*

A higher level of integration in the system supports the system's ability to come up with an appropriate responsive behavior of the system as a whole in relation to gravity. In whatever circumstances the body as a whole might be situated, it has more "possible possibilities"—more options to chose from—for solving the equation of being a human being in constant relation to gravity. It is not only that we might find better solutions that are more efficient and appropriate, but it is also the case that we have different ways to solve the same equation or problem. If we relate this to the already mentioned concept of robustness, we then have more redundancy and diversity in the system, and this creates more robustness in the system.[10]

The body as a whole system can be developed to a higher order of function and complexity

A premise for the concept of integration is that it is possible to develop the body as a whole system. This development is something we could call kinesthetic development. In our case, we are interested in certain aspects of kinesthetic development, namely how the body as a whole system, with its intrasystemic elements, interacts with gravity. There are other ways of studying kinesthetic development that do not have to directly involve the body's relationship to gravity: for example, looking at the development of the complexity in motor control, or looking at the development of strength and speed, and so forth. One thing to remember though is that we often confuse adaptation of the whole system with the development of the whole system. When the system adapts, it increases the function of the system in certain narrow circumstances, with a trade-off in diversity and functionality in other circumstances. That is, we pay by a lack of diversity and functionality when it comes to other non-adapted activities. For example, an avid marathon runner's metabolic rate might become lower at the times when they are not running in an attempt to conserve energy. Alternatively, a mixed martial arts fighter might get very good at squeezing somebody tight between their arms, but this happens at the expense of losing some of the efficiency when doing activities that require total flexion in the shoulder, like changing a light bulb in the ceiling.

In contrast, when the system as a whole develops it increases the complexity and diversity of the system as a whole so that there is a greater opportunity for enhanced function in whatever circumstance the system might be situated. There is no trade-off but a general enhancement of the system as a whole. In order to be the best elite athlete, having one's system maximally developed and maximally adapted is probably a necessity. However, usually, we put more efforts into adapting the system than developing the system. A marathon runner, for example, usually puts more effort into perfecting the training regime than into looking at how the interaction and dynamics of the intrasystemic elements could be enhanced to better support efficiency in the system as a whole.

We usually have no problem with understanding that children need to develop certain aspects of themselves before they are considered grown up. When it comes to learning mathematics, for example, we have a sense that it is not something that develops on its own; there is a need to go to school and study for it to happen. But when it comes to our kinesthetic development in the sense that we have been talking about above, there seems to be a lack of awareness that such a thing exists. Developing the body as a whole system in relation to gravity is looked at as something that is supposed to happen by itself. Because gravity is not something that is possible to escape, the child hopefully learns to stand up, walk, and move about in some way. Nevertheless, we might be better off if we acknowledged that kinesthetic development and the body as a whole in relation to gravity is of importance.

It is my view that the kinesthetic development related to the body as a whole and its relation to gravity has a deep relationship to other forms of development and other systems within the body. If we take an example for comparison, moral development is dependent on cognitive development. If we do not have the capacity to take another person's perspective cognitively, it might be hard to understand the impacts our actions might or might not have on other people. In the same way, our emotional development might be dependent on our kinesthetic development. If we lack basic kinesthetic development, we might have a more difficult time at self-regulating our emotional status. It is not difficult to see that robustness, integrity, and integration in one system can affect another system, or even that integrity and integration in one system is a necessary condition for integrity and integration in another. If we look at interaction from a socio-psychological viewpoint, we can see that limits in our kinesthetic development might put restrictions on our ability to grow into different social roles. What if our kinesthetic development is a necessary and intrinsic part in our overall development? Given our lack of attention to this at a societal level, where does this leave us?

Summary of the principles

Principle 1: Language and concepts come with underlying assumptions and premises.

Principle 2: The body is a system of systems within complex systems.

Principle 3: This system can be developed to a higher order of function and complexity.

Principle 4: The body as a system and its relation to gravity (where gait is included) is an integral and necessary part of kinesthetic development.

Notes

1. I am well aware that the reductionist paradigm as it is portrayed here might not be a homogeneous coherent paradigm as such, and that it is a bit simplified and undiversified to match the narrative in this article.
2. For further inquiry into this, see Lindberg (1992). Science seems to have taken a distinct, more analytical turn with Descartes. Stuart Kauffman and Arran Gare (2017) take a similar starting point with Descartes (and Newton), and from there develop a view that is distinctively more open and less deterministic than what usually is the norm.
3. The need to include subjectivity in physics, or "Endophysics," and the fact that we cannot get out of our own way, is discussed in Kauffman and Gare (2017).
4. It seems as though this was the case in the early days of molecular biology (Crick 1970). The point here is to describe which level of organization is causally potent and active, the causal linearity, and the degree of complexity.
5. Mittra (2007).

6. Chemotherapy is not my area of expertise, and the intention here is simply to give an example, rather than to discuss the effectiveness or ineffectiveness of chemotherapy.
7. For a discussion on different levels of organization and their causal efficacy, see, for example, Dupré (2008).
8. Kitano (2004).
9. Again this is not my area of expertise, and there may exist other studies that contradict these studies; the point is to show what can constitute an example, not the level of "truth" in the claims that the study is making (Gerstein et al. 2008).
10. Kitano (2004).

References

Crick FHC; Central dogma of molecular biology. *Nature* 1970, 227:561–563. DOI:10.1038/227561a0.

Dupré J; *The Disorder of Things: Metaphysical Foundations of the Disunity of Science.* Harvard University Press, 2008.

Gerstein HC, Miller ME, Byington RP, Goff DC Jr, Bigger JT, Buse JB, et al.; Effects of intensive glucose lowering in type 2 diabetes. *New England Journal of Medicine* 2008, 358:2545–2559. DOI: 10.1056/NEJMoa0802743.

Kauffman S, Gare A; Beyond Descartes and Newton: recovering life and humanity. *Progress in Biophysics and Molecular Biology* 2017, 119(3):219–244.

Kitano H; Biological robustness. Nature Publishing Group. *Nature Reviews Genetics* 2004, 5(11):826–837.

Lindberg DC; *The Beginnings of Western Science.* University of Chicago Press, 1992.

Mittra I; The disconnection between tumor response and survival. *Nature Clinical Practice Oncology* 2007, 4(4):203.

CHAPTER 5

Relationships

Linus Johansson

You know that good relationships can be something absolutely fantastic. A beautiful relationship with someone you love or a creative and prosperous relationship with a coworker and colleague is one of the greatest things that can happen in life. Good social relationships are also a direct and very strong indicator to a long and healthy life.[1] You also know that a bad relationship can be something very difficult and demanding. It can wear you down and cause you a lot of pain and suffering. Relationships are also virtually impossible to avoid. The fact is that if you are, to some extent, coexisting with another person—be it lover, coworker, child, neighbor—you have a relationship. It is the quality of your relationship with the other person that makes you two function or not function together.

A relationship is dependent on several aspects if it is to be healthy and prosperous. One of the most crucial is communication. How you form your words into meanings, wishes, or descriptions for the other party to perceive and relate to will reinforce or shatter the foundation of your relationship. The other important aspect is integrity. Although you have a relationship with another person, the need to be able to be an individual and to have the space and freedom to do what you please without the other party disturbing or interfering with it is also very important for the health of the relationship.

All this said, we cannot underestimate the importance of relationships. They can make an immense difference in how you feel and perceive your life. However, relationships do not only exist outside the body, they also reside within the human form and structure. From the perspective of this book, seeing and understanding relationships within the human body is one of the more important key elements in successfully helping another person develop the ability to move and function through gravity. It is one of the most important keystones in movement integration.

Just as the two main aspects for an external relationship, communication and integrity, play a great role for the outcome of the perception of your life, so too for the relationships in the human form.

With communication in eternal relationships we refer to the ability for a structure to have a good proprioceptive function. With integrity we refer to the fact that a structure should do only what it is "designed" to do and not be compelled to do or handle movement or forces that it was not designed for.

To give a simplified example, when we move it is most important for the foot to have the ability to move as a foot should and to have good proprioceptive communication with the rest of the body via the nervous system and ultimately the brain. This will allow us to successfully accomplish the movements we wish for. However, problems can occur when the foot loses its ability to move as a foot should, for example, due to injury or unsuitable footwear. When this happens the foot can no longer do its proper task in the movement chain. This leads not only to incorrect movement but also incorrect proprioceptive information supplied to the rest of the body.

The first thing that happens when someone tries to walk or run using a dysfunctional foot like this is that the level of communication drops or alters due to the altered movement. There will be less sliding and gliding over joint surfaces and less tension and loading of tissues surrounding and interacting with the foot. The bottom line is that the foot becomes more "quiet" with less proprioception flowing to the brain in combination with the lack of movement. However, the brain still tries to be successful in completing the desired task of walking or running and creates a compensation, as we have already mentioned in Chapter 2.

The goal of compensation is to be as successful as possible by distributing the movement that should have happened in the foot somewhere else in the body. In this theoretical example, the brain chooses the lower back as the escape route and forces that structure to do more movement and handle greater forces than it is used to, so as to be able to create the desired task. This soon leads to a pain pattern occurring in the lower back region due to an overload of movement and movement forces.

What has happened in this example is also what happens in so many other relationships. Communication is failing and the lack of understanding leads to bad decisions which in turn leads to someone doing something that they did not have the capacity to do. In an external relationship between two people this might lead to an argument or a verbal fight. In the body it almost always leads to pain.

The similarities between ordinary external relationships between two people and the relationships in the human form are quite remarkable when we choose to see them this way. It is quite obvious that they can be described with the same principles. Our mission, in a situation like this, is to use our interventions to create increased communication between the two parties and to sort out the tasks between them, thus creating an optimal composition of relationships in a person's body that gives them the perception of lightness and less hindrance when moving through gravity, and even the sensation of less pain.

However, when working with clients it soon becomes obvious that there is no unified definition of how a person should be, structurally or functionally. Instead of stating an exact description of how a body should be, we therefore use the expression "optimal body," defined as the body that suits the person's needs and wishes and, foremost, a body that fits their lifestyle.

Not everyone needs to be able to run fast or deadlift heavy weights; for some people it is just getting up in the morning and reaching the jar on the top shelf that is the optimal life. The optimal body is therefore a variable definition and each person has their own unique optimal body to strive for.

The concept of the word "strive" is essential in this perspective. Creating a goal for the person to reach is one thing; however, it is not when that goal is reached that change happens in the body. Rather, it is while they are striving toward their goal that the changes and developments take place. This means that we can set goals for a client, but we must understand that we gain development during the period of striving for the goal. One could argue that the striving is, in itself, the goal.

This way of perceiving and understanding relationships within a human structure and the possibilities this gives to develop a person's abilities to move and function in gravity is truly a creative endeavor.

Creating relationships

There are many relationships surrounding the human being and we need to acknowledge them all to be able to make progress when working with the human form. We can put them into three major categories, and we do that with very fuzzy lines. First, there are all the external cognitive and emotional relationships with other people and places that we create by just living. Then we have our body's external relationship to gravity and the surfaces that we move over, i.e. the earth. This relationship was given to us all when we were brought into this world by our mothers. Finally, we have all the internal relationships between all the structures in the body. These internal relationships are formed and developed in the interaction of the two other categories of relationships, the cognitive and emotional and the dynamic relation to gravity and the earth.

Before moving further, please note that we always acknowledge that all aspects of the human form and function are connected, integrated, and dependent on each other, and this is especially true when it comes to the three categories of relationships. They all depend deeply on each other and if one of them alters the others will change too. This is something very powerful to know and understand as it means we can work with one relationship in the human form to affect and alter a completely different relationship elsewhere. All practitioners working with the human mind know this. Changing something in a person's mind can alter how they perceive, appreciate, and foremost, move their body. The great thing about this is that it also works the other way around. Movement can play a part in helping a person suffering from depression, for example.

Seeing it this way we can also understand that these categories of relationships are very much a part of who we are and how we function, in body and mind. We all depend deeply on them and we know that we must appreciate them all to be able to create an interpretation of our true function.

We cover our view of the cognitive and emotional relationships in Chapter 12. In this part of the book we give you our interpretation of internal relationships and how they create the interface for the external relationship with gravity and the surfaces we move over.

We believe that one has to see and understand internal relationships first to be able to understand how the human form functions in relation to gravity and the more complex and functional movements of real life. We begin this by sorting out all the various concepts to state our perspective on this.

A classic relationship that has been around for a long time is that of the human body in relation to vertical and horizontal lines. Vertical and horizontal lines are two constants: a string makes a vertical line when a weight is hung at its end, pointing to the center of the planet. From that vertical line the horizontal line is created and extends at 90° from the vertical line, according to the Pythagorean theorem.

These two perfect and much needed lines are used when constructing a building. Without these lines we would not, for example, get straight walls and flat floors. They are therefore a crucial necessity when constructing solid structures that will stand for hundreds of years. In many practices they are also used to assess the human

structure, mainly to define posture. The structures are set in relation to the vertical or horizontal line and assessed accordingly. An intervention is applied to correct any deviation from the lines.

From our perspective we do not see the human form as a standing and solid structure that can be set in relation to these two constants. We see the human body as a free, fluid, and organic form, impossible to relate to any fixed lines. A building is supposed to stand still and never move, while the human body is supposed to do quite the opposite. Through hundreds of thousands of years of evolution our body has been chiseled to do nothing else but to always be in motion, never to be still. For that reason alone we, in this perspective, will never lock the body up within the boundaries of any constructed constants.

The question then becomes obvious—when we say that we work with "relationships" and do not relate to horizontal and vertical lines, or any other outside constants, what then do we create relationships to? The relationships we see and acknowledge are within the human form and ultimately with the dynamic interaction with gravity and the surface that we move over.

One might argue that the representation of gravity is the vertical line in this scenario, but we beg to differ. The classic vertical line is drawn straight through the middle of the body as one unified constant derived from the idea of the weight at the end of a string. Gravity, however, works on each and every cell in the human body. One cannot unify the force of gravity to be a centered, dead, and constant force working only in the center of a fluid and organic form. The effect of gravity on the human form varies depending on the internal, external, and even emotional relationships that reside within each and every human form.

By default we also say that gravity is a pulling force. That might be true for the weight at the end of the string; however, we are not suspended from a string, but move over the face of the earth on our two feet. In many respects we could argue that gravity is not pulling us down but instead is pushing us up. Each time our feet touch the ground we do not fall through the ground, but instead get "pushed up," keeping us going and moving. In some sciences this is interpreted as "the ground reaction force."

As all forms of knowledge are only interpretations of the perception of the world around us, we are free to turn the game, if needed, and see things the other way around, especially if this gives us a fresh perspective and helps us to better help our clients reach their desired goals.

To fully understand and appreciate this we begin by explaining it from the concept of internal relationships before connecting to gravity.

The principle of relating the structures to each other when assessing the body render the use of the vertical and horizontal lines unnecessary, especially when you assess in positions and movements that are closer to life. This will probably give more "true" findings, at least from our perspective.

This way of perceiving and understanding relationships within a human structure and the possibilities this gives to develop a person's abilities to move and function in gravity is truly a creative endeavor. We will unravel this concept further throughout this chapter.

To be able to work with the internal relationships of the body one must have a principle and a given and stated language to use to be able to document the findings and to be able to communicate the finding to others. A language that is very useful is that presented by Michael Morrison in his article "A structural vocabulary"[2] and made famous in the book *Anatomy Trains*[3] written by Tom Myers. In this language four different movement terms are used to describe any given occurred event in the body. They are: tilt, shift, rotate, and bend.

In the more classic therapeutic world, the use of static findings and fixed positions are more common to describe what is being observed, e.g. genu varum, genu valgum, lordosis, kyphosis, scoliosis, eversion, inversion, etc.

The cunning aspect of using movement as a descriptor of what is happening in a body is that it is dynamic and also that it will guide the observer to what is needed to be done to create a change and development.

If stating that a structure has moved in a certain way the logical solution will be to guide and facilitate the structure to also be able to move in the other direction, to create a more optimal relationship and strive for the optimal body.

We will go deeper into this principle further into the book.

Tensegrity

"If it's somewhere, it is everywhere."

To further understand and see the complexity of relationships and integrity inside and outside of the body we can use the so-called "tensegrity model."

The word "tensegrity" is derived from the combination of two words, "tensional" and "integrity." The concept originated in the 1960s with Buckminster Fuller, designer, architect, and inventor. The tensegrity model is a concept used mainly for constructions and buildings. It is a structural principle based on the use of components that are isolated via compression inside a net of tension. The most common models to exemplify this are constructed out of struts connected via rubber bands. Due to the way the struts and bands are connected the components do not touch each other but create a floating construction. This construction keeps the form by tension and integrity and can deform when being loaded with external forces, only to regain its form when released.

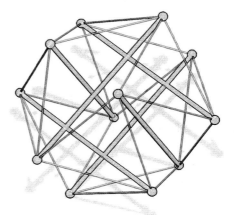

5.1

All the struts are held in isolated compression in relation to each other and create a kind of floating construction.

The tensegrity model has also been widely used in the more holistic approach to address some of the properties of human function. As the tensegrity model is a very simple construction, with limited function compared to the human form, it must be used with caution when presented as an analogy. With this in mind we can still use several of the properties in the tensegrity model to get a simplified insight into human complexity and the idea that it is "all connected."

To convert this to our analogy of relationships, we can also see the tensions that the rubber bands create between the struts as "communication" and the fact that everything is floating and not touching we can see as the "integrity," much as in the human form.

When applying an external force to one or several of the struts in the model, we can see that the tension this creates is distributed all through the entire structure and all components are involved in handling this. This is very much as when the human form moves through gravity and forces are applied and loading the system.

Another way to affect the entire tensegrity structure is if we shorten some of the elastic bands that connect to the struts. The direct deformation and altered relationship will obviously occur in our hand; however, we can also see how this local deformation transforms, and moves further into the tensegrity model to affect the entire structure, creating altered relationships between the struts and affecting the tension in the rubber band between them.

It is clear that this deformation will take away much of the structure's original form and shape; however, what is interesting for us is that it will also affect the ability to equally distribute any applied external load throughout the structure. The bands that are held short will no longer yield, elongate, or communicate as external forces are applied to the model; this will in turn lead to more movement being distributed over the other rubber bands and putting greater tension into them. This simplified example can also be seen in the human structure and an unequal distribution of movement-induced force like this could be argued to be an underlying reason why damage to structures appears.

We can see this shortening of the rubber bands as the foot in the example given earlier. The foot had stopped moving and communicating as it was supposed to and the movement forces had to be distributed somewhere else, creating dysfunction or even pain.

However, there is a counterpoint to this; something interesting happens when we do the opposite. Instead of crumpling the rubber bands and creating shortening, we take two of the struts, one in each hand, and we separate them. This will, instead of compromising, give more volume and space to the entire structure as a whole. What is remarkable is that we see the same thing in the human structure. When given more space, freedom, and openness in one place the body will not compromise and shorten somewhere else, instead this will ripple like waves on water and create more openness and freedom elsewhere.

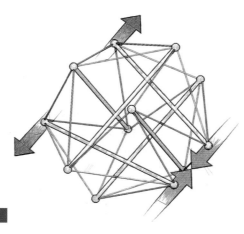

5.2

In this figure we can see that due to the compression of the two struts' ends by the red arrows we get an equally unbalanced opening between the two other ends of the struts.

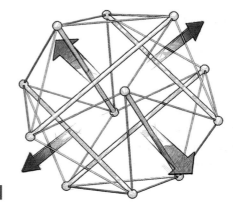

5.3

When an even and balanced opening is applied to the tensegrity model it create openness in the entire structure.

As for the example with the foot, if we can reintegrate the movement and communication via the foot back to the entire system and work to move away from the compensations the person can start to distribute movement more equally through the body. The sensation will be more openness and freedom and the experience of increased ability to move and do more demanding exercises. This of course also applies to people without a specific movement disorder that just want to develop and increase their movement potential. If we see and understand this we can use it very successfully when working with clients. It opens up many possibilities and foremost it lets us focus on the person as an entire system, able to see beyond the idea of "bits and pieces."

We acknowledge that the body is infinitely more complex than a tensegrity model; however, the same simple and basic principles apply and teach us that all is connected and that, if it's somewhere, it's everywhere.

Interrelative movement

> *Hey look, it has moved!*
> *OK, in relation to what?*

We can simply conclude that movement is a major key to everything that regards function and health in the human body. Therefore it is of the utmost importance that we understand what movement is and how it is defined.

> *"To describe the movement of an object it has to be done in relation to another still or moving object."*

From this definition we can divide movement into three subcategories; absolute movement, relative movement and interrelative movement. Absolute movement is described as the movement of an object set in relation to a constant reference point that does not move, i.e. a spot on earth. Relative movement is described as the movement of an object in relation to another object, which in turn also has the potential to move in relation to a constant reference point, i.e. a spot on earth.

For example, you stand with a friend. You throw a ball in one direction and your friend starts to run in the

opposite direction (figure 5.4A). The spot on earth that you stand on is now the absolute reference point. If we describe each movement separately in relation to that spot, we describe two absolute movements. The distance the ball covered (B) and the distance your friend covered (A), during the same time frame. However, if we describe the movement that occurred between the ball and your friend, in relation to each other, we have instead described one relative movement (X).

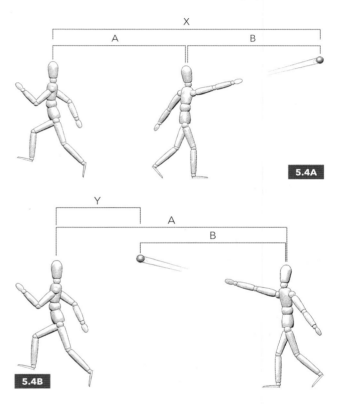

When absolute movement is described it does not make any difference in what direction you throw the ball and your friend runs, it will always be described as the same distance in relation to the given spot. However, in terms of relative movement the direction plays a great role. If your friend runs in the opposite direction the relative distance will increase. If your friend runs in the same direction as you throw the ball the relative distance will decrease. In figure 5.4B we see that the relative distance Y has decreased due to the change of the direction the ball was thrown, compared to X in figure 5.4A.

What can seem to be a movement occurring in one way or a certain distance, in absolute space, can, in relation to another structure, show no movement at all, or even a movement in the opposite direction.

To add yet another deepened definition of movement we can also speak of the "interrelative" movement between two objects. This interrelative movement does not take into account absolute space or any other reference point. Interrelative movement only describes the relationship between two given objects or structures with respect to each other. In this sense it does not matter which object we choose as a reference, it is still the same movement that has occurred.

Imagine two spaceships, a red one and a blue one, in an absolutely empty space. You can be in either one of them, looking out at the other, but you cannot be outside observing both as a third reference point. Standing on the red spaceship looking at the blue you suddenly experience the blue spaceship moving away from you.

In this instance, the only thing you can say is that the space between you and the other spaceship increases. Whether it is the blue spaceship moving away from you, or you on the red spaceship moving away from the blue, or both spaceships moving away from each other, it is impossible to say as there is no absolute space or third reference point.

From the perspective of interrelative movement, it is unimportant which object does which movement, as we don't include any reference point to absolute space, only the altered relationship between the objects is important. If we would like to make things clearer and express a higher degree of complexity we can include a third object and describe the relationship and movement between all these three objects.

We will dive deeper into this example further in this chapter.

This is quite important to grasp and understand, as the interrelative movement is a key factor to the concept of loading tissues eccentrically. The eccentric load is in turn the foundation to create energy efficient movement. This is mentioned in Chapter 3 and explained further in Chapter 6.

If we go back to the principles of movement we can conclude that, although we have two principles to use to describe movement, it seems that absolute movement always becomes the default principle used when describing movement of an object. The reasons for this can be several. However, the main one is probably because absolute movement is what we get taught in our different educations and schools. It is a logical and simple principle to use and it is easy to teach.

Seen from our perspective, however, absolute movement has obvious limitations and is not as potent as the principle of interrelative movement when it comes to understanding and working with the human form, and we will soon explain why.

Our experience is that absolute movement is firmly fixed in the way we perceive and explain the world around us, sometimes to the degree that we are totally oblivious to the fact that we cannot break free from that way of thinking. To see and understand this we first have to touch back on what we have mentioned earlier, namely the vertical and horizontal lines, and how they relate to absolute movement. First, start by looking around where you are right now. If you are in a man made environment and not sitting on a stone deep in a forest, you can probably see that the vertical and horizontal line is present more or less everywhere, right?

The walls around you are straight like the vertical line. The floor beneath you is parallel to the horizontal plane.

5.5

The doorframe has two vertical lines and a horizontal. The table in the room has a surface perfectly parallel to the horizontal line. Everything around you that we have constructed, and that in some manner is weight-bearing, is arranged according to these two lines. Even if you see a beam in the roof or along a wall that has an angle or tilt, that angle is always a result of Pythagorean theorem and is the product of one vertical and one horizontal line in a triangle.

Our problem is that we cannot escape this subconscious input that we are constantly fed with, especially because gravity plays an active and very strong factor in all this. If things are not straight and squared to each other, they keep falling, tipping over, and rolling away all the time, creating chaos around us. We can't have that! We therefore become experts on creating the most stable relationships between objects through gravity. We stack, lean, bind, and solidify objects in relation to the vertical and horizontal lines so they become as static as possible and do not create chaos. This is quite the opposite of the free and fluid body we live in.

We are extremely affected by this way of being and we allow our mind to deceive us all the time, seeing these lines everywhere and referencing them to everything. Even the book you are reading right now has these lines. The fold in the middle of the book subconsciously becomes the vertical line and the top and bottom of the page become the horizontal lines. The lines of the text become horizontal. Even the pictures with a colored background have a 90° relation in each corner and a very good art director has made everything in this book, text and pictures, all line up very neatly according to these two lines. You would soon become confused and upset and you would not be able to concentrate if the book was completely disordered and the text and pictures were tilted in all different directions. Right?

However, there is a paradox to all this and for us it is a light in the end of the tunnel. Regardless of how, subconsciously, we align everything to these lines, we also very much appreciate organic forms that break away from them. For example, imagine a room with four walls, several rectangular windows, and a door. In the middle of the room stands a table, squared to the walls and with a surface parallel with the floor. In the middle of the table stands a vase and in this

vase is a lush bouquet of flowers. The flowers do not line up to any horizontal or vertical line—they are an irregular and organic form in the middle of a sea of straight lines. Which makes them extremely appealing to behold. Somehow we are torn between these two worlds. It is obvious that to be able to make constructions and explain absolute relationships in room and space, we are dependent on these two lines; however, when appreciating natural beauty we are more drawn to organic forms that are free from these constraints.

Nonetheless, we are all a part of this squared world and we know most of us will have a hard time leaving behind this way of seeing things, especially when it comes to the human form. We copy and paste the concept of the vertical and horizontal line into our work with the body, even when we all know that we are more related to a lush bouquet of flowers than to a room with four walls.

You perhaps didn't even know there was another way of seeing it. That is why it is going to be interesting to see if we can make you see what we see and help you break free from these two constructed lines.

It is important not to entirely put aside or dismiss the importance of the idea of the vertical and horizontal lines or the concept of absolute movement. They are much needed in many other things we do. The key is instead to be able also to learn to see, understand, and appreciate interrelative movement and to think about how we can use that knowledge to develop our clients' movement potential. The importance is therefore to fully understand the principle of interrelative movement and the potential it brings.

We will give another example to take this to the next level. You stand in a room with a friend by a table. On the table are two cubes, one red and one gray (figure 5.6A1). They are positioned in relation to the edge of the table. The red cube is 1 cm from the edge and the gray cube is 20 cm from the edge.

You step out of the room and your friend rearranges the two cubes. You step back into the room and your task is now to describe how the cubes have been rearranged on the table (figure 5.6A2).

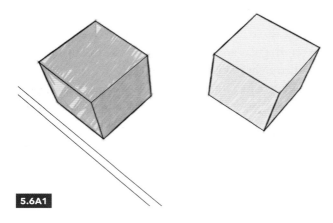

5.6A1

You find that the red cube has been moved further away from the edge of the table by 5 cm and the gray cube has been moved closer to the edge of the table by 5 cm. When telling your friend what has happened you have described two actions of absolute movement, one for each cube in relation to the same reference point, i.e. the edge of the table.

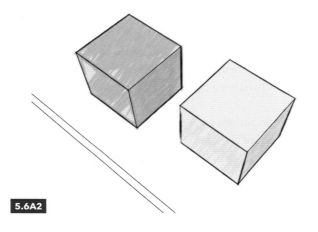

5.6A2

The interrelative movement that has occurred in this scenario, on the other hand, is that the two cubes have moved closer to each other by 10 cm. In relation to the edge of the table they have moved 5 cm each, but in relation to each other, they have traveled 10 cm.

The most important aspect about interrelative movement—and this is necessary to understand and remember when we are working with relationships in the body—is that if both cubes had been moved an equal distance and in an equal direction there would have been no interrelative movement between them. The two cubes would have made an absolute movement in the room in relation to the edge but not in relation

to each other. This is something that can create a lot of confusion if it is not understood.

This phenomenon is why the use of interrelative movement is much more interesting and powerful to use when working with human movement, and we will keep explaining why.

Let's take this even further and remove the third reference point, to more clearly see the idea of interrelative movement. You stand with your friend in the same room, this time with an infinite table, which means that the table has no edges, i.e. there is no absolute reference point. On the table are the same two cubes. They stand next to each other with 10 cm between them. You once again leave the room for a minute to let your friend rearrange the two cubes.

You step back into the room to once again describe how the cubes have been rearranged. Remember, as there are no edges on the table you can't use them as reference, you only have the cubes in relation to each other and must describe that altered relationship using interrelative movement.

What you can clearly see is that they have been moved apart by approximately 10 cm more and that there are now around 20 cm between them. The question is now, has the red cube been moved away from the gray cube, has the gray cube been moved away from the red, or have they both been moved in each direction to create the new relationship (see figure 5.6B1-3)?

As you do not have an absolute reference point to relate the cubes to you do not know. And that, in itself, is the answer: you don't know, and you don't have to know! You just have to see and be able to describe the new relationship between the two cubes.

Now, the power of interrelative movement is not to understand how the relationship has altered as much as to understand what we must do to bring the relationship back. In this scenario we describe the relationship the cubes had before you left the room as the optimal relationship, and that is the relationship we want to bring back.

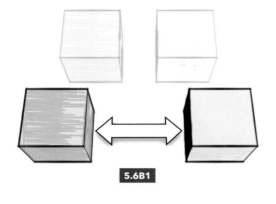

5.6B1

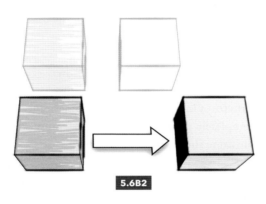

5.6B2

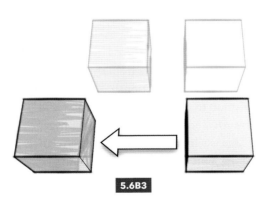

5.6B3

The same goes when working with our clients. When we see an altered relationship between structures we can describe this state and how we want to work to bring it back. The way we see and describe it also becomes the indirect description to how we can work to bring the optimal relationships back.

Back to our cubes. To be able to rearrange them back to the desired relationship we first have to choose how to describe the movement that has occurred between them, then we can choose an intervention to bring the

relationship back. To do this we first have to choose one of the cubes to be the reference point for the second cube's movement, in order to be able to describe the movement.

Now, what is very important to understand when doing this is that you also accept that the two other options of movement also have occurred, at the same time. You have chosen one but the two others still apply.

This is important, because when working with the human form, we sometimes have to realize that the first option of movement we choose will not have as much impact on creating change as we had hoped. We can then go back and choose another of the options of movement.

Hence, nothing is lost or done wrongly, we just have to make an adjustment in how we see and describe the interrelative movement between the two objects and alter our intervention accordingly. In comparison, if we use absolute movement to describe an altered state we cannot go back and make any adjustments, as there are no other options.

How to pick the reference point?

We can pick any structure in the body as a reference point and describe all other structures in relation to that point. However, there is also a logic to picking a reference point and the more you work from this perspective the more logical it gets as it also intertwines very closely with the interventions.

When we are looking at a person and we see altered relationships between structures, we certainly also see structures that have not altered. These structures *seem* to be more in an optimal state to each other. Picking one of them as the reference point is therefore the most logical thing to do.

The best way to describe this is to add another cube to our infinite table. Now we have a red, a gray, and a blue cube in a row on the table, still with no edges on the table. We decide that figure 5.6C is the optimal relationship between the cubes.

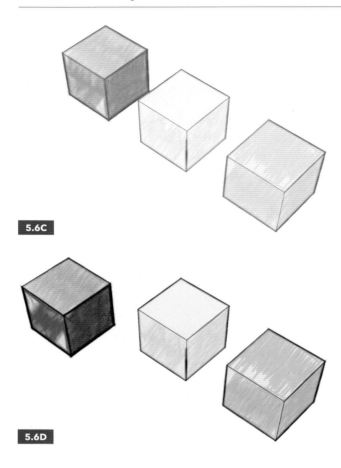

5.6C

5.6D

In figure 5.6D we can see that the relationship has altered. We can, like in the previous example with the red and gray cube, describe the three different scenarios of movement to have happened between the cubes. However, the first thing that comes to mind is that the blue cube has left the optimal relationship to the gray and red cube. The gray and red have the same relationship as before. Although one can never be sure, it could be that the gray and red have moved away from the blue and still be in relation. However, when

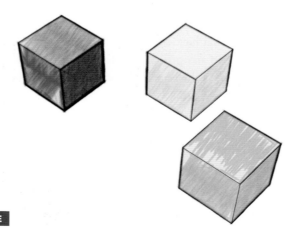

5.6E

choosing an intervention it would be most logical, time and energy-efficient to first try and bring the blue cube back to the optimal state. This becomes even more evident when working with a person.

In figure 5.6E we see yet another altered relationship. Is it the gray cube that has moved, the red and blue that have moved, or all three? We don't know, but once again it seems more logical to try and move the gray cube to regain the optimal relationship.

Bear in mind that this is simple when working with cubes. Working with structures in a human body is something completely different and far more complex. However, the same simple principles apply.

5.6F

If we look at a person, as in figure 5.6F, and we see that the chest and the feet are in a relative optimal relation to each other the most logical description would be that the pelvis has moved forward and away from that relationship. This makes it easier to choose an intervention to try and bring the relationship back. However, if we do not get results from the intervention based on the first described relationship we can quickly alter the description to another reference point and alter the intervention accordingly and hopefully get a better result. This is why the principle of interrelative movement is a dynamic way to work as it always presents many solutions to altered relationships in a human structure.

Needless to say, up to this point we have just worked with the structural relationships. We have yet to learn and understand how this principle also applies to movement. Although we see and describe everything as movement, we do it in static positions. Yet another dimension opens up when we later put the structures to the test and see if they can actually move or not.

This will in turn be a part of an entire treatment plan where many relationships are addressed and altered and where the ultimate relationship to address finally will be that of the entire movement system in relation to the gravitational forces of the earth. All this will hopefully bring the person back to a state of being able to experience the sensation of moving freely without any hindrance or pain.

Please once again note that there is not one generic optimal structural relationship that all humans should be fitted too. Each person has their own optimal state of relationships. The only way to learn and see this is to start assessing and treating people, using this perspective, and it soon will become obvious how each individual's optimal version should be. An individual's optimal state of relationships is always built up from a combination of the person's wishes, abilities, and prerequisites fused together with the current skills and knowledge that we as practitioners bring to the table.

Leaving the cubes and focusing on the human structure, we can quickly conclude that there are more than just a few simple structures to take into account. The human body is replete with complex relationships and structures and they are all connected, making it a vast and very sophisticated maze of variables.

Unlike the two cubes, the body's composition and fundamental function through gravity lays out some obvious rules for integrated relationships. In the example with the cubes you could have easily stacked them on top of each other without any hindrance. As for the human body, all structures are bound in their actual relationships and they are "designed" to follow the rhythm of the gait pattern. This fact creates a kind of playground for us to stay within and it makes our job both more logical and at the same time more complex and demanding.

The movement of structures

To be able to see and appreciate movement we must be able to describe movement. As mentioned before, we use the four movement terms when describing what has happened in the relationships within the human form. We call this structural assessment. Used correctly and adequately we can make up a "story" to describe the current relationships residing within the person's form. From that we can then create the foundation to choose our interventions and rewrite the story. This will help us clear a "path of solution" to reveal the best and

most efficient way of developing the person to suit their wishes and needs.

These four movement terms are in themselves simple; however, when used to describe relative movements in the body they can explain very complex relationships that would have been more or less impossible to describe using "old school" terminology.

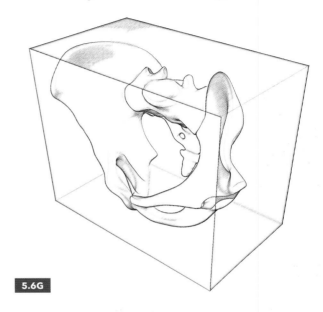
5.6G

To be able to use these movement terms together with the structures in the body we first need to turn each structure into a theoretical box. A box has six sides; however, we only use the front and the top side when describing the orientation of a movement. What is the front and top side of the structure-box is determined by how we would generically describe the front and top of the human body. We also use the person's left and right when describing movements to the sides, and lateral and medial to describe the movements away from and closer to the middle of the body.

At this point, to describe and clarify the properties of each

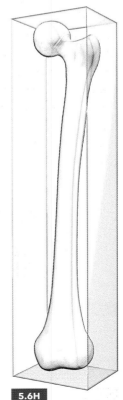
5.6H

movement term, we will illustrate and explain them in relationship to the prior position they had in space and not in relation to another structure.

The four movement terms are:

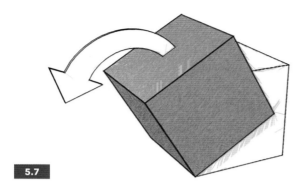

5.7

+ Tilt—when a structure moves into a sloping position, rotating around an axis in the frontal or sagittal plane.

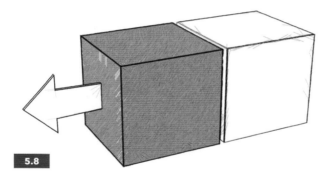

5.8

+ Shift—when a structure moves or changes from one position or direction to another in the frontal or sagittal plane.

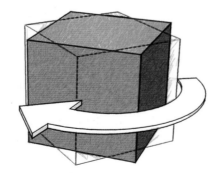

5.9

+ Rotate—when a structure turns around an axis in the transverse plane.

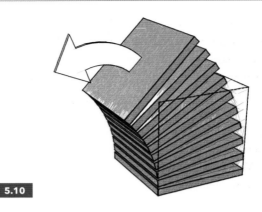

5.10

+ Bend—when a curve is created in a structure or when a series of structures have tilted in a sequence to create a curve, e.g. vertebrae.

A classic example

We will now give a few examples of altered relations in the body and how they can be described using interrelative movement. We have constructed the following examples to let you challenge yourself and not relate the movements we describe to the horizontal and vertical line, i.e absolute space. Let's see if we can broaden your perspective.

Tilt in the sagittal plane

The first example is perhaps the most common relationship we see when assessing our clients. It is the relative tilt of the pelvis and ribcage in the frontal plane.

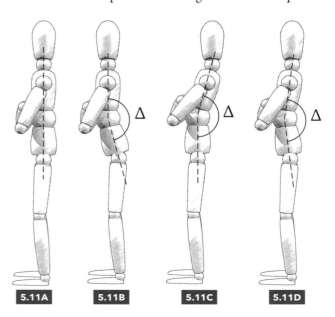

5.11A   **5.11B**   **5.11C**   **5.11D**

In this example we describe figure 5.11A as showing the optimal relationship between the feet, pelvis, chest, and head. Looking at the three other figures 5.11B, C, and D, you can clearly see that all three have quite different altered positions. Or do they?

In this example the relation between the pelvis and the chest is actually the same in all the altered positions, B, C and D. We describe it as the delta angle, $\Delta°$, which is the amount of degrees that has changed between the pelvis and the chest in relation to the optimal position, A. Even if the "posture" of the dummy looks very different in the three scenarios the relationship between pelvis and chest is constant.

Even though the relationship between the pelvis and chest is constant throughout this example, how they in turn relate to the rest of the body will give us a great variation of relationships and lead us to a great variation of interventions.

Structural assessment

Looking at figure 5.11B we can clearly see the altered position between pelvis and chest marked as delta angle, $\Delta°$. To describe the internal relationships to the rest of the body we need, as mentioned earlier, to first choose our reference point to describe what has happened.

When looking at figure 5.11B, our eyes are immediately drawn to the pelvis and the tilt that has occurred. If we were to make the feet our reference point we would describe the pelvis to have tilted forward in relation to the feet and not the vertical line, as it does not exist in this perspective.

However, when looking at the chest and the feet, we see they do not have an altered relationship when it comes to tilt in the sagittal plane. What has happened instead is that the chest has shifted forward in relation to the feet in the sagittal plane.

Moving on to figure 5.11C, we can see that it has the exact same relation between pelvis and chest as figure 5.11B; however, in this scenario the pelvis is neutral to the feet. If we instead pick the pelvis as the reference point we can describe the chest to be tilted backward in relation

to the pelvis. The head in turn is neutral to the pelvis and feet when it comes to tilt but the head has shifted backward in relation to the pelvis.

Moving on to the last figure 5.11D, just as for the two dummies before, the relationship between chest and pelvis is the same. What has happened here is that we have a mix of figures 5.11B and 5.11C. What we can see first is that the feet and head have a neutral relationship, no tilt or shift has occurred in the sagittal plane. If we therefore once again pick the feet as reference point we can describe the pelvis to be tilted forward in relation to the feet and the chest can be described as tilted backward in relation to the feet.

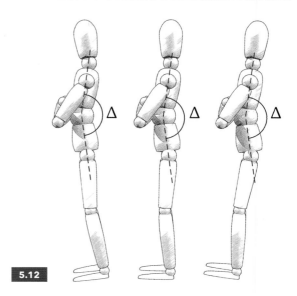

5.12

Just to show that the relationship between the pelvis and chest has been the same in all three dummies, we have put the three figures together with the chest and pelvis all in neutral relation to each other (figure 5.12). Please note that each figure has its previous internal relationships intact. From this compiled figure everything seems to be thrown out of order and this is only because the eye tries to align it to the vertical line.

We have just focused on the relationship between the tilt of the pelvis and chest and the way we can see and describe it in relation to all other structures. Let us now go back and look at figures 5.11B, C, and D, and to the head, and describe its relationships more closely. In all the illustrations the head would commonly be described as "neutral", in relation to the vertical line. Once again, this has nothing to do with the vertical or horizontal line—this is the position the head usually takes when a person is asked to look straight ahead. Our eyes and sensory organs in our ears help us coordinate the head to a position where we can observe and perceive the world around us in the most optimal way and this is almost always done regardless of the ongoing relationships in the rest of the body.

If we look at figure 5.11B and the relationship between the head and the feet there is a clear shift between them but no tilt. If we look at the relationship between the head and the chest it is neutral, with no tilt or shift. However, if we look at the relationship between the head and the pelvis the head is tilted backward in relation to the pelvis, and so is the chest. This is important to see and understand. Why? Because if we were to only alter the position of the pelvis in relation to the feet and not also address the chest and neck the person would be leaning backward, i.e. be positioned as figure 5.11C. Therefore, as much as we need the pelvis to tilt backward in figure 5.11B, we also need the chest and head to tilt forward in relation to the pelvis if we want to get to the position of figure 5.11A.

In figure 5.11C this becomes even more obvious. Here we can clearly see that the head is tilted forward in relation to the chest. If we were to only alter the chest's relationship with the pelvis and tilt it forward, and not include the head's relationship to the chest, the dummy would be standing looking down.

In real life this would never happen. A person would never be standing looking down at the floor after an applied intervention. What they probably would do, subconsciously, is to not accept the altered relationship between chest and pelvis and instead return to the previous position. They could also perhaps alter the position of the head themselves, depending on how deep that relationship is. However, not until we can see and understand that we also need to address more than one relationship—in this case also between the chest and head—can we get the results we are looking for. To get the chest to tilt forward in relation to the pelvis we also need the head to tilt backward in relation to the chest. The head's position to the rest of the body is crucial to address if we ever want to get any change when working with a person. The reason for this is

that people always try to hold the head in a neutral relationship to the rest of the world and however the relationship has altered beneath the chin, the head always stays more or less afloat with the rest of the world.

Hence, if we were to only relate the head to a vertical line we would probably never see the need to create altered relationships with the rest of the body. No one would say that the head needed to tilt backward or forward if it was in perfect alignment with the vertical line. That is why interrelative movement to the internal structures in the body is a very powerful way of seeing and working, especially in this perspective.

What we have done here is to make some quite lengthy conclusions. In fact, we do not actually know if the relationships we see are true until we have put them to the test. What we do when we do a structural assessment like this is to raise suspicions. These suspicions are then tested, to be either confirmed or dismissed. If both the structural assessment and the test show that there is a potential for development, then we will take that on board and compile it into a treatment plan.

Rotations in the transverse plane

To make this more understandable we deliberately explain one movement plane at a time. But in real life there are no movement planes: all motion always happens in all planes at the same time. We only reduce it down to each plane to make it understandable at this point. Further along in this book we will put it all together into a more connected explanatory model.

This fact is very obvious when working with the transverse plane because the altered relationships we see in this example are in many cases products of altered relationships in the sagittal and frontal planes. Hence, if we first work with what we find in the two other planes we might not have to address any altered relationships in the transverse plane.

But for now, in this example, we address the transverse plane and the properties in this example.

Structural assessment

Just as in the sagittal plane and in the earlier example, we need to look at the structures and choose our reference point to describe what has happened, then lay down the idea of what we would like to do for this person to help them find balance and a higher level of function.

To aid your learning process we have deliberately made this example so that you can instantly start to refer all structures to the horizontal line by adding the dotted lines on head, chest, pelvis, and feet. Let's see if you can break loose from that way of thinking and see what we see.

Now, what is interesting is that what has truly altered in this relationship is not the dummy, it is you. The relationships between the head/chest and pelvis/feet are exactly the same in figure 5.13B and C. The only thing that has altered is you. In figure 5.13B we have a neutral relationship to the head/chest when looking at it and would therefore describe the pelvis/feet to be altered in this relationship. And then, for some magical reason, we change our mind when looking at figure 5.13C. Now we are in neutral to the pelvis/feet and all of a sudden the altered relationship is the head/chest.

What you need to understand is that your relationship with a client also affects the relationships you perceive. Both these descriptions are therefore correct and both altered relationships exist at the same time. We just need to decide which we want to work with and why.

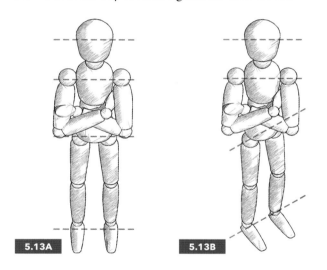

5.13A   5.13B

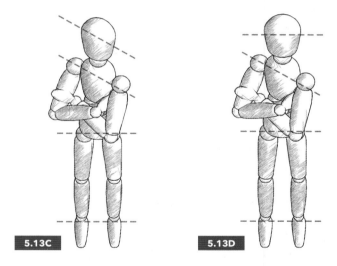

5.13C   5.13D

Let's start with figure 5.13A and call that the optimal state and desired relationships.

Looking at figure 5.13B we would immediately say that the pelvis and feet have turned to the left in relation to the chest. Why we say that is only because the chest and head are in a "neutral" in relation to us looking at the person. (We always describe using the person's left and right.)

Looking at figure 5.13C we would instead say that the chest and head have turned to the right in relation to the pelvis and feet. In this figure we are instead in neutral with the pelvis and the feet.

And this brings us to figure 5.13D. Now, what has happened here? It differs from figure 5.13C in one aspect; the head also is in neutral in relation to the pelvis and feet. The chest is now the structure that is most altered from its "optimal relation." We have in this case several structures that are dominant in their relationship and one that has altered. This would then lead us to say that the chest is rotated to the right in relation to the head, pelvis, and feet. However it would not be wrong to say that the head, pelvis, and feet have all rotated to the left in relation to the chest.

In conclusion

Relationships are everywhere, internally and externally, and depending on how we perceive and describe them we can form very creative and dynamic interventions to alter and optimize these relationships for the client's best outcome.

In Chapter 19, Julian Baker speaks of the relationships that different fascial structures have in the body. Due to the individual qualities each structure has, they provide, through their relationships, fantastic potential for our abilities to move and function.

In this chapter we have spoken of the relationships in a more static and structural sense. Altered, enhanced, and more dynamic movement is what we are striving for and this is the theoretical explanation that provides support for our principles and further down the line, our interventions and methods.

We will in the next chapter, Oppositions, dive even deeper into the world of relationships and find the more functional aspects of it all.

Notes

1. Holt-Lunstad (2018).
2. Morrison (2001).
3. Myers (2014).

References

Holt-Lunstad J; Why social relationships are important for physical health: a systems approach to understanding and modifying risk and protection. *Annual Review of Psychology* 2018, 69:437–458.

Morrison M; *Structural Vocabulary—A Plain English Vocabulary for Describing Geometric Relationships.* Boulder, CO: Rolf institute, Rolf lines, July 2001.

Myers TW; *Anatomy Trains: Myofascial Meridians for Manual and Movement Therapists, 3rd edn.* Edinburgh: Churchill Livingstone, 2014.

CHAPTER 6

Oppositions

Linus Johansson

All relationships in the human structure are important for our function and ability to move. However, we would like to argue that there are two relationships in the body that perhaps are more interesting than others. Interesting because they are the two core relationships that all other movement relations in the body depend on to flourish and be successful. Due to the unique properties that reside within these two relationships one can also claim them to be the very foundation of the human gait.

These two relationships are the relation between the forefoot and the rear foot and the relation between the pelvis and the chest.

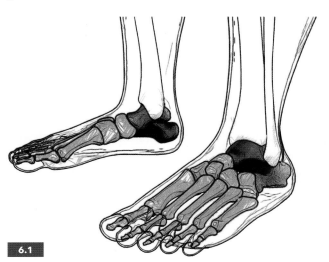

6.1

When we refer to the rear foot we speak of the talus and calcaneus and when we speak of the forefoot we refer to all the metatarsal bones. However, there is no border dividing the foot into these parts and the definition can vary if needed.

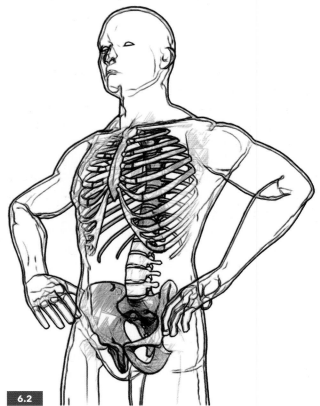

6.2

The pelvis is seen as the two ilium halves, the sacrum, and a fading portion of the lower spine. The ribcage is seen as the thoracic spine fading out into the cervical spine and lower back, and the ribs connecting into the sternum. The shoulder girdle moves in relation to the ribcage and is therefore seen as a free entity.

As explained in Chapter 10, there are no borders between structures; it is all one. Therefore we must acknowledge that the structures we speak of can only be truly seen through function. All those structures that we name as "parts" must be viewed with very blurred lines and we must acknowledge that these imaginary lines are only for explanatory and educational purposes.

When we describe these two unique relationships, the one between forefoot and rear foot and the one between pelvis and chest, we say that they are in a relationship of opposition toward each other.

To be in opposition in a relationship is defined as to always do the opposite to what the other part of the relationship is doing. That means that when we see a structure in a relationship of opposition move one way, we want to be able to describe the other part of the relationship to move in the interrelative opposite direction.

When looking at the foot in function, ideally we want to see a clear opposition of movement between the forefoot and rear foot. In undisturbed function they should always reflect each other's movements and when they do we can describe that as the compiled movement of pronation and resupination.

The same thing happens in the relationship between the pelvis and the ribcage, in function. Ideally we want to be able to describe both structures to move in each others interrelative and opposite direction. (This is described in greater detail in Chapter 8.)

We will soon explain in more detail what the "purpose" of the opposing movements is and how they contribute to the complexity of human function. But first we need to once again step back and make some important conclusions before moving on.

We know that we can see and explain each of the movement terms tilt, shift, rotate, and bend by themselves in any of the three different movement planes, sagittal, frontal, and transverse. This is done purely for educational purposes, to be able to start to understand each one of the movements and the properties they have. In reality, none of the movements ever happens by itself in one single plane of movement when we move through gravity—they always happen at the same time in all three planes.

Why we present the movements one by one in each movement plane is simply because our minds lack the ability to overview big and complex processes. It is impossible for us to see and explain all things that happen in human function at the same time. We do not have that capability. We need to "segregate" to learn, but more importantly, we also need to learn not to stay segregated.

Segregation is therefore acceptable to some degree; however, we must, as soon as we have seen and understood one of the movements in one of the planes, step back and anchor it to the holistic reality. We will not get the profound response or result we seek if we do not treat the body like the indivisible and intelligent unit it is.

In Chapter 3, we spoke of the theories behind human evolution and that we are the only mammals to habitually walk on two legs. The main reason for us getting up on two legs was probably to become more energy efficient. The function of opposing movements in the body plays a big part in this energy efficiency.

To actively contract a muscle to create movement demands a lot of energy and is a very inefficient way of moving. Have you tried moving just by contracting your muscles? You will move like a robot and completely without any flow or smoothness.

The paradox is that although muscles can contract in a concentric fashion, it is the opposite motion that creates energy-efficient movements and flow. In the human movement system the tissues are arranged in such a way that they are able to create something called elastic recoil. Robert Schleip gives a full account of elastic recoil and the properties around it in his book, *Fascial Fitness*.[1]

In contrast to the concentric shortening of a muscle, where structures are moved closer together, elastic recoil happens when a muscle is loaded eccentrically through a movement where structures are moved further apart in a controlled movement through gravity. Due to the elastic properties of the tissue it recoils back and creates a passive contraction and a shortening between the structures again. This means that we can load our movement system through gravity using the body's own weight and passively receive a movement of recoil with very little effort and energy. Gary Carter elaborates more on this phenomenon in Chapter 17.

You can try this right now to feel the difference. Stand up on your two legs and start to do concentric calf raises. Hold on to the back of a chair if you need support. Do 5, 10, 15, 20 … Soon you begin to feel the calf muscles in your lower leg. Imagine walking and running and feeling this. You wouldn't get far creating movement in that way. Now rest for a minute or two.

This time, instead of contracting your calves and still with straight legs, bounce on the balls of your feet. Can you feel that there is no great feeling of contraction in the calf muscles, but instead they have turned out to feel more like rubber bands. When you land on the ball of the foot the calves elongate eccentrically and before the heel touches the floor they recoil and bounce you back up again. Now you can do 20, 30, 40 bounces or more without even getting tired. Can you feel that this is more the feeling you have when you walk and run and that you have become much more energy efficient?

In a study from 2002,[2] Kawakami et al. showed that during an eccentrically loaded counter-movement plantar flexion of the foot the muscular part of the calves stayed more or less isometrical and did not elongate, whereas the fascial elements in the calf elongated and contributed to the movement. You hopefully noticed this when you did the same thing bouncing.

The example given here was for the calf muscles; however, this phenomenon happens in all structures associated with human movement. To be able to create an eccentric movement and load the tissue, the parts where the tissues reach between must be separated to create an eccentric movement. This is one of the key features of opposing movement. Instead of having one structure being still and the other moving the needed distance, they each move in opposite directions to create the distance needed and sufficient elongation of the tissue.

As the function of opposing movements provides the optimal basis for elongation of tissue to create elastic

recoil, it also creates the optimal dynamic and structural internal relations to interact with gravity.

When we land on one foot in the gait we do not need to shift our entire upper body over to one side to maintain balance. Instead we can internally organize our structures in a way that makes our center of gravity project over our base of support, i.e. the foot.

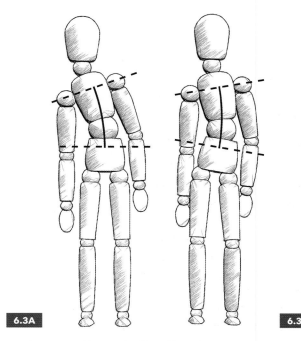

6.3A   6.3B

Try this yourself. Stand tall and bend to one side (figure 6.3A). Feel the effort needed to get the elongation of the tissues on the outside of the body doing this. Now try what the dummy in figure 6.3B is doing by bending one knee and tilt the pelvis first and then tilt the chest in the opposite direction. Can you feel the elongation on the outside of the torso and with much less effort?

In this figure 6.3A we see how the dummy keeps the pelvis neutral to the feet and tilts the chest to create the desired elongation between the lower parts of the ribcage and upper parts of the ilium. In figure 6.3B, however, the dummy tilts the pelvis halfway in one direction and the chest halfway in the opposite direction, creating equal elongation to the dummy in figure 6.3A without moving off center and losing the center of support.

In figure 6.3A we can understand that more body weight is placed on the right foot due to the quite hefty tilt of the upper body. In figure 6.3B the dummy has instead bent the left knee and lets the pelvis tilt to the left and the chest to the right, keeping more or less the same amount of weight on both feet.

What else is evident is that the dummy in figure 6.3A will soon get a sore neck due to the compensatory tilt back to the left, in relation to the chest. In figure 6.3B the compensatory tilt of the head is not as demanding as the chest has not tilted as much as in figure 6.3A. The head's relationship to the chest in figure 6.3B therefore clearly shows that the principle of opposition is a more efficient distribution of movement. The compensatory movement of the head is less in figure 6.3B than in figure 6.3A, and less compensation is always more energy efficient.

In figure 6.3B the dummy had to bend the left knee to be able to let the pelvis tilt to the left. This means that more movement gets connected when the principle of opposition is at play.

This is even more evident in the foot. When the foot moves as it is "designed" to do in the gait, cycling through the movements of pronation and resupination, it drives a biomechanical rhythm, via the talus, up the body, and connects to the opposing movements in the torso. Though, depending on how we want to see and express things, we can also say that the rest of the body, via the talus, drives the foot through the cycles of pronation and resupination. All this will be explained in more detail in Chapter 8.

In conclusion

We see opposition in human movement as the deeply integrated function that amplifies the biomechanical rhythm of the gait and contributes to the optimal creation of elastic recoil in the tissues. Via elastic recoil and the ability to keep us centered over our very narrow base of support when moving, the function of opposition is one of the fundamental pieces that makes us the highly energy-efficient movers that we are.

We, in turn, use these correlations to assess, test, and develop a person's abilities to move and function when moving through gravity.

Moving further into this book, and deeper into the explanatory models, we will soon set all of this in a more dynamic relationship to the gait.

In Chapter 8, we invite our evolutionary heritage into the principles of relationships and present yet another level of seeing, perceiving and understanding human function and movement.

Notes

1. Schleip and Bayer (2017).
2. Kawakami et al. (2002).

References

Kawakami Y, Muraoka T, Ito S, Kanehisa H, Fukunaga T; In vivo muscle fibre behaviour during counter-movement exercise in humans reveals a significant role for tendon elasticity. *Journal of Physiology* 2002, 540(2):635–646. doi: 10.1113/jphysiol.2001.013459.

Schleip R, Bayer J; *Facial Fitness: How to be Vital, Elastic and Dynamic in Everyday Life and Sport*. Chichester: Lotus, 2017.

CHAPTER 7

The Color Illustration Model: a new way to illustrate movement

Martin Lundgren

Explaining movement and complex phenomena in the body only using language can be a daunting task. If we were to use only language to describe interrelative movement, it would make things unnecessarily hard and also quite boring to read. The need for some form of illustration is therefore quite obvious. As usually mentioned, a picture says more than a thousand words. The intention when forming this way to illustrate interrelative movement has been to make it as clear as possible and minimize any ambiguity and confusion. A further requirement has been that it can show a high degree of complexity without making things very complicated.

The standard way to show some form of movement in a static picture is to put an arrow over a skeletal part to indicate that it has moved or is moving in some way, as, for example, in figure 7.1. The unexpressed assumption in this way to illustrate is that the skeletal

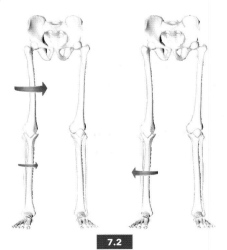

part moves in relation to absolute space itself, as we talked about in the last chapter. This is something that is very ingrained in our way of thinking, and it seems like it is our unconscious default position when we start to think about movement. If we want to make things a bit more complex, we usually start to talk about changes in angles in a joint or difference in speed of the bones when they move in a joint. If, for example, the tibia also internally rotates like the

7.1

femur did in figure 7.1 but does so to a lesser degree and not as fast as the femur, it is commonly pointed out that we get a relative external rotation of the tibia. In figure 7.2 the tibia in the skeleton on the left side does not rotate as much or as fast as the femur, which means that we get a relative external rotation of the tibia, like in the skeleton on the right side. But when we say that the femur rotates "more" or "faster" it does so in relation to a reference point of absolute space itself. Again the unexpressed premise here is that the bones move in relation to absolute space itself. This relative view of movement stands in contrast to the interrelative view which does not presuppose a reference point to absolute space. In the relative view we have concluded that there is something that is forward and backward, left and right, and so forth, like an invisible grid that the bones move in relation to. This absolute space is usually represented by a Cartesian coordinate system, as in figure 7.3.

7.2

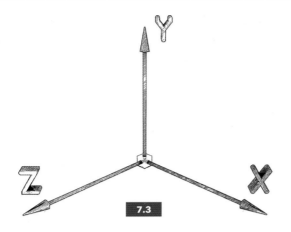

7.3

The only problem is that in the real world there isn't a grid, there is no forward or backward other than what we have decided. Usually what is forward is where the nose or eyes are pointing, or perhaps where our feet are pointing, which way we are traveling or from the viewpoint of the observer. We could, of course, create a grid with computer analysis or draw a grid on the wall when we do a structural assessment, but this misses the point. This is why when we talk about interrelative movement we only talk about one bone's movement in relation to some other bone, and there is no comparison to absolute space itself. This way to illustrate primarily describes intrasystemic movement relationships or, put in different words, how different elements within the body move in relation to each other. There are many reasons for this: first, this way of looking at interrelative movement makes it a lot easier to go from assessment to intervention. Second, if we want to increase the complexity in the standard way of looking at and describing movement, things can turn complicated very fast.

In comparison, this intrasystemic interrelative view on movement that we represent, can handle complexity very well without making things overly complicated. This is not to say that the standard biomechanical way to describe movement does not have any value, it is just that it is less applicable in this context. In some circumstances, there is a need to relate the system as a whole or a part of the system to something outside itself, i.e. the earth itself. But this is then clearly pointed out and follows the same logic as when we relate movement within the body itself.

We have chosen primarily to look at how the bones move in the body. You could choose to look at other structures or "parts" in the body first, but we have concluded that the most logical starting point when looking at movement is to start to look at how the bones move. Once we understand how the bones move, or if you want, "float" in the flesh of the body, one can start to look at other constraining factors in the body, infer loading patterns, look at soft tissue qualities, or other aspects that are relevant for movement. When we say that we look at how the bones move in the body, we look at how one bone as a whole moves in relation to another bone, more than what happens locally in the joint. In the interrelative view of movement the bone or selection of bones that another bone moves in relation to determines the reference point and which way is "forward".

The Color Illustration Model

To illustrate this way of looking at interrelative movement I have created the Color Illustration Model. This creation will hopefully make it a bit easier to understand complex movement patterns in the body without complicating things too much. As the name implies, what a given movement of the skeletal part is moving in relation to, is illustrated by a color. For example, in figure 7.4 the femur is rotating internally in relation to the pelvis; this is represented by the color red in the arrow in front of the femur. The same movement can be illustrated with the pelvis as a reference point; in figure 7.5 we have a yellow arrow in front of the pelvis pointing to the left, indicating that the pelvis is rotating toward the left in relation to the femur. This is a way to illustrate the same movement; it does not matter if we look at the red or the yellow arrow, it is still the same movement (from a standpoint of interrelative movement).

On the other hand, the femur's or the tibia's movement might be very different to something else in the body. If we want to make things a bit more complex, we have to relate the movement to something else in the body. If we look at figure 7.5 again, there is no movement between the femur and the tibia; the femur and the tibia are pointing the same way. If we look at figure 7.6, there is still the same movement happening between the femur and the pelvis, but if we look at the relationship between the femur and the tibia we can see that the femur is internally rotating in relation to the tibia as well, represented by the blue arrow. So the femur is rotating internally in relation to both the pelvis and the tibia. In figure 7.7 we can see that we have the same movement between the femur and the pelvis as we had in figure 7.5, but this time the relationship between the pelvis and the ribcage is different. We can now see that the pelvis is

rotating to the left in relation to both the femur and the ribcage. So in short, to get a more complete picture of what is happening in the body, we might need to relate the movement to more than one bone in the body (or a selection of bones).

The size of the arrows indicates how much movement is happening, a small arrow indicates less movement than a big arrow. The arrows in the pictures describe a movement if not otherwise stated. We need to remember that describing a movement is very different from describing a static position. Describing a static

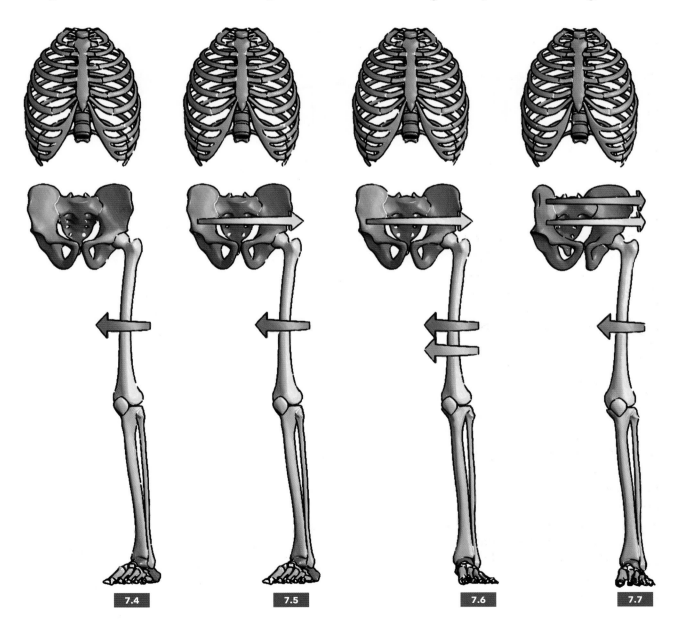

7.4   7.5   7.6   7.7

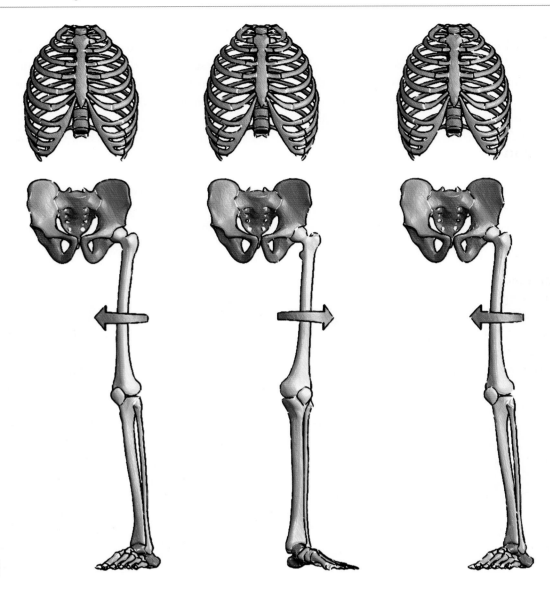

7.8

position does not entail movement in any way. Imagine that the skeleton on the left side in figure 7.8 describes a static position. The femur is rotated internally in relation to the pelvis. It could be that the femur is stuck in this position. This would mean that we do not get any movement of external rotation of the femur in relation to the pelvis like the skeleton in the middle. Neither would we get any movement of internal rotation of the femur in relation to the pelvis like in the skeleton on the right. For this to happen the femur would have to be more neutral or externally rotated in

relation to the pelvis first; otherwise, the femur would not have any distance to travel because it is already at its end range. The opposite could also be true, that the femur is positioned in an internally rotated position in relation to the pelvis, but it is not "stuck" in this position, and when we move the femur, the movement of internal and external rotation in relation to the pelvis is fully available. Hence the need for some form of assessment that involves movement and not only position when we work with the body.

CHAPTER 8

Gait made simple

Martin Lundgren

Introduction

As already mentioned throughout this book, when it comes to being a human being, gait is one of our primary essential movement patterns. The purpose of Gait Made Simple (GMS) is for you to be able to see where people are and where you can take them, for you to be able to see how you can develop people's kinesthetic ability. People are different from each other, and their gait is different. Still, there are some essential movements that we want to see when people walk. To know what moves in relation to something else is crucial for creating strategies when we form different kinds of interventions. GMS is not only a way to learn how to do gait assessments, but it is also a way to get a better sense of movement in general. Furthermore, it is also an introduction into interrelative movement and a way to understand the dynamics in the body from a systemic perspective. Moreover, it is also a way to clarify relationships in the body that otherwise would be hard to identify and understand. The intention with Gait Made Simple is to present a simplified system so we can go easier from assessment to intervention, so we know what to do with whom, and where to go next.

Background, premises, and the concept of opposition

One of the underlying assumptions or ideas behind GMS is that of opposition between pelvis and ribcage and between rear foot and forefoot. The movement of opposition simply means that two things move in opposite directions in relation to each other, as in figure 8.1. Gary Ward in his Anatomy in Motion courses presented this concept to me in 2011. The movement of opposition is also an underlying theme for his Flow Motion Model of Gait™. This concept of opposition has served me well and has worked as a good entry point in the conceptual terrain when trying to explain how I have been developing my work over the years. This is why this is the basis for Gait Made Simple. The intention with GMS is for you to get an understanding of the essential movements in gait and how they relate to each other. It aims at keeping as much complexity as possible without overcomplicating things too much. The Color Illustration Model is an intrinsic and essential part of GMS. To correlate GMS to kinematics studies and other relevant research is beyond the scope of this pedagogical context. That endeavor would be more suitable in another form, a more extended advanced form of Gait Made Simple.[1]

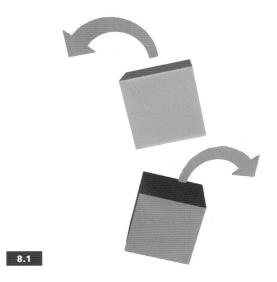

8.1

I will describe different movements one plane of motion at a time; this is to make things more understandable, not to say that all these movements only happens one plane of motion at a time. In the body, things, of course, move in all planes of motion at the same time (though sometimes more in one than in another).

Different assessment strategies when doing gait assessment

When we do gait assessment, we can take advantage of three different assessment strategies:

1. *Looking at asymmetry between the sides.* Do we have the same amount of movement on both sides? Do we, for example, have the same amount of movement when it comes to pelvic tilts in the frontal plane on both sides? For example, in figure 8.2 we have more tilt toward the left in the pelvis in relation to the ribcage in the skeleton on the left than we have right tilt in the pelvis in relation to the ribcage in the other skeleton.

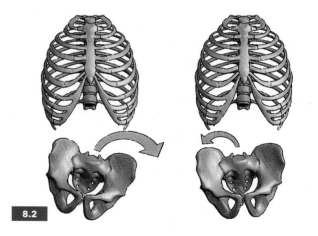

8.2

2. *Estimating how much movement we have in general for a particular segment, part, or bone.* For example, we can look at how much movement of pelvic tilt we have in the frontal plane. We might lack the movement of pelvic tilt on both sides, or we might have a lot of movement on both sides. In figure 8.3 we have a lot of movement in the pelvis both when it tilts to the left and when it tilts to the right in relation to the ribcage. And in figure 8.4 we have

a situation where we have very little movement in the frontal plane in the pelvis both when it tilts to the left and when it tilts the right in relation to the ribcage.

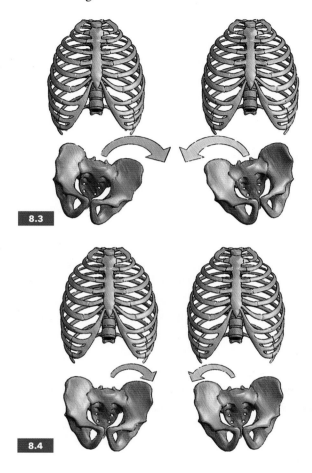

8.3

8.4

3. *Examining the relationship between the movement of different bones, and seeing how that relates to movement elsewhere in the body.* A lack or excess of movement in one place might affect movement somewhere else. If we, for example, lack movement between the talus and the tibia that might affect the amount of anterior tilt we get in the femur in relation to the pelvis. If we look at the skeleton on the left side in figure 8.5, there is less anterior tilt in the femur in relation to the pelvis compared to the skeleton on the right, because of what happens between the talus and the tibia. In the skeleton on the right side, we get more anterior tilt in the femur in relation to the pelvis because we have no movement restriction between the talus and the tibia.

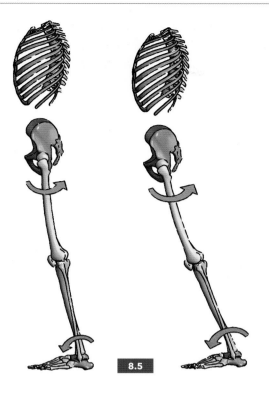

8.5

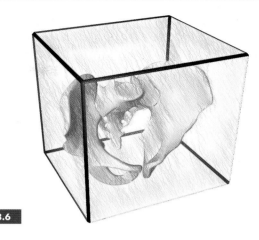

8.6

Strategy 1 is usually easier to look at because we have a direct contrast between the sides. In strategy 2 we have to contrast the amount of movement we see in the person with a more abstract notion of how much movement is possible or appropriate. This comparison is usually a bit harder to see and takes some more practice because we have to develop a reference point. Strategy 3 is more of an analytical undertaking where we need some kind of a system as a guide. The intention with this Gait Made Simple is to be such a guide.

Movements of the pelvis in gait

We are going to start looking at how the pelvis moves in gait. We will treat the pelvis as one whole unit, with no intrapelvic movement; for simplicity's sake we can visualize the shape of the pelvis when it moves as if it were like a box (figure 8.6). This is so we can get a sense of the overall global movement of the pelvis. We are going to start by looking at the movement of the pelvis and relate it to the ribcage and the femurs.

Pelvic tilt in the frontal plane

Every time the foot is put forward and weight is put on it, there is going to be a tilt in the pelvis. So if the right foot is put forward, there is going to be a tilt toward the left (figure 8.7). And if the left foot is put forward and it hits the ground, there is going to be a tilt toward the right. In figure 8.7, the pelvis tilts to the left in relation to both the ribcage and the femur, represented by the green and yellow arrows.

Pelvic shift in the frontal plane

The pelvis also shifts in the frontal plane. It moves as a whole from side to side. When we swing one leg forward, the pelvis shifts from being over the foot on the back leg to being over the foot on the front leg. Looking at figure 8.8, the pelvis in the skeleton on the left side is more over the left foot; the pelvis then shifts, and in the skeleton on the right side the pelvis is more over the right foot. This shift is represented by the orange arrow, which shows

8.7

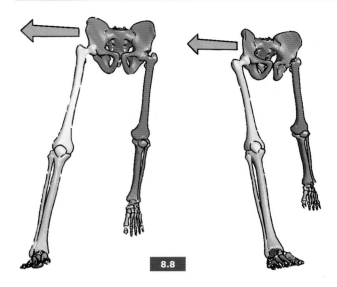

8.8

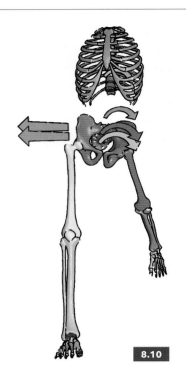

8.10

that the pelvis shifts to the right in relation to the feet. The pelvis also shifts in relation to the ribcage. If we look at figure 8.9, there is also a green arrow just next to the pelvis, demonstrating that the pelvis is shifting toward the right in relation to the ribcage. There is also a pink and yellow arrow over the pelvis indicating that the pelvis is tilting in relation to the femurs. But there is no green arrow indicating a tilting movement, as in figure 8.7. This absence is because the pelvis is not tilting in relation to the ribcage. In relation to the femurs, on the other hand, the pelvis is tilting, as a shift of the pelvis in relation to the feet also necessitates a tilting movement in relation to the femurs.

8.9

As long as the foot of the swinging leg is in the air, we do not have any tilting relationship between the ribcage and the pelvis. However, when the whole of the foot hits the ground, the shifting movement and the tilting movement of the pelvis are joined together, as in figure 8.10. The shifting of the pelvis starts earlier than the tilting, but they end up together.

Pelvic rotation in the transverse plane

Rotations of the pelvis occur in the transverse plane. Every time we put one foot forward the pelvis rotates away from the front leg. If we put the right leg forward, the pelvis rotates toward the left in relation to both the femur and the ribcage (figure 8.11). This movement comes to its maximum just before the foot of the back leg comes off the ground. It is quite common to see a big difference between the sides when we look at this movement. A hint can be that when we see someone walk, the bellybutton turns one way, but never crosses the mid-line to turn the other way (strategy 1).

8.11

Anterior and posterior tilt in the sagittal plane

In the sagittal plane, the pelvis moves in an anterior and posterior tilt. Every time the front leg is put forward and the whole foot hits the ground the pelvis tilts anteriorly. In figure 8.12 the pelvis tilts anteriorly both in relation to the femur and the ribcage. The anterior tilting is a way for the body to shock absorb and load up the soft tissue. This movement ends when the back leg starts to swing forward, which starts to drive the pelvis posteriorly (figure 8.13). This posterior tilting continues until the foot hits the ground and the pelvis again goes into anterior tilt.

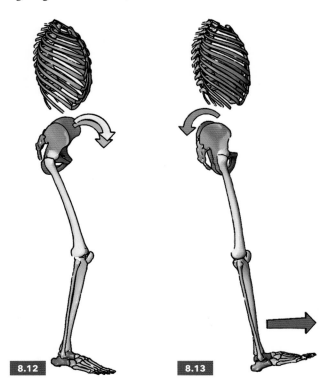

8.12   8.13

This movement can be a bit hard to see because we usually look at a person walking toward us or away from us. Still, this pelvic movement when the pelvis goes from posterior tilt to anterior tilt is very significant, especially for creating posterior bending up the spine. If we compare the amount of anterior tilt that happens when one foot hits the ground compared to the other, we usually do not see much of a difference. In the same way, if we compare the amount of posterior tilt that takes place when swinging one leg forward compared to the other, we also usually don't see very much of a difference (strategy 1). That is because it is much more common to see that a person is either stuck in a posteriorly or anteriorly tilted position. The pelvis stays anterior the whole time and does not posteriorly tilt when we swing the leg, or it stays posterior and does not anterior tilt when the whole of the foot hits the ground (strategy 2).

The relationship between the ribcage and the pelvis

When we looked at the pelvis, we looked at it as if it were one unit, with no movement inside of itself. We are now going to treat the ribcage the same, as an undivided whole, as if it were a box. The ribcage is of course not a box, and there is plenty of movement going on between different elements in the ribcage. However, as we said before, we do this so we can understand the overall global patterns of movement in the body. The easiest way to understand what happens in the ribcage is to say that it does the exact opposite to the pelvis. We have an opposition between the pelvis and the ribcage, seen from an interrelative perspective. We are now going to explore the dynamics between the ribcage and the pelvis.

Movements in the frontal plane

If we have opposition between the pelvis and the ribcage, the ribcage is going to move away in relation to the pelvis in the frontal plane. If we put the right leg forward and the pelvis tilts to the left, the ribcage is going to tilt to the right in relation to the pelvis, as in figure 8.14. If we relate the tilting of the ribcage to something else, like the right femur, for example, we can see that the ribcage also tilts in relation to the femur (figure 8.15). The tilting and shifting of the pelvis, explained earlier, helps to balance the weight of the person over the right foot. If there is less of this movement we have to counteract it somehow; one typical example of this is that we get more of a tilt in the ribcage, because nothing is happening in the pelvis. If we explain this in terms of interrelative movement and look at figure 8.16 we can see that the ribcage is tilting more to the right in relation to the femur than it did in figure 8.15, and less in relation to the pelvis. In some rare cases the pelvis even tilts the opposite way to what we have described here; the ribcage then has to tilt even more in relation to the femur to compensate for this.

As we mentioned earlier, the pelvis shifts in the frontal plane. When the pelvis shifts from the back leg to the front leg the ribcage remains more centered over the back leg. When the ribcage stays behind in this way, this means that we will get a shift between the pelvis and the ribcage. As the right leg swings forward to hit the ground, the pelvis shifts to the right in relation to

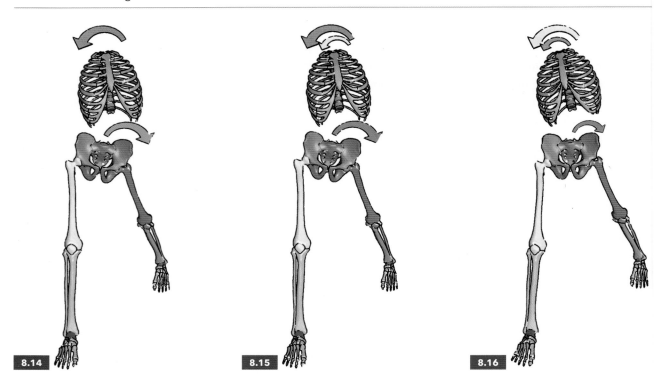

8.14   8.15   8.16

the ribcage (figure 8.17). This is a way for us to move from being centered over the foot on the back leg to preparing for being centered over the foot on the front leg. The shifting occurs primarily when we have one foot on the ground. As soon as the whole of the foot in the front leg hits the ground the ribcage catches up, the shift between ribcage and pelvis diminishes, and we instead start to get more of a tilting relationship between pelvis and ribcage, as in figure 8.14.

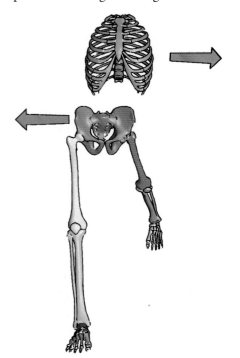

8.17

Looking at shifts between the ribcage and pelvis can be very revealing. It is common that we have quite a big difference between the capacity of the ribcage to shift left versus right in relation to the pelvis. If we see a significant shift in the frontal plane between the ribcage and the pelvis in a static visual assessment it is also quite common to see the same shift when we look at the person walking. Imagine that figure 8.18A is a static position in a visual assessment; there is a shift between the pelvis and ribcage (enhanced for visual purposes). It is then common to see that shift enhanced in gait; in figure 8.18B, the ribcage is shifting to the left in relation to the pelvis when we are swinging the right leg forward. When we swing the left leg forward, on the other hand, as in figure 8.18C, there is no shift at all between the ribcage and pelvis. If we do a movement assessment, asking the person to keep the ribcage horizontal and then shift the pelvis and ribcage in relation to each other, we will probably notice that it is easier to shift toward the side that the person stands shifted toward, and toward the side the person shifts when they walk.

A good thing to remember is that a shifting relationship between the ribcage and the pelvis usually involves two bends in the spine, but the tilting relationship between the pelvis and ribcage involves one bend (and maybe

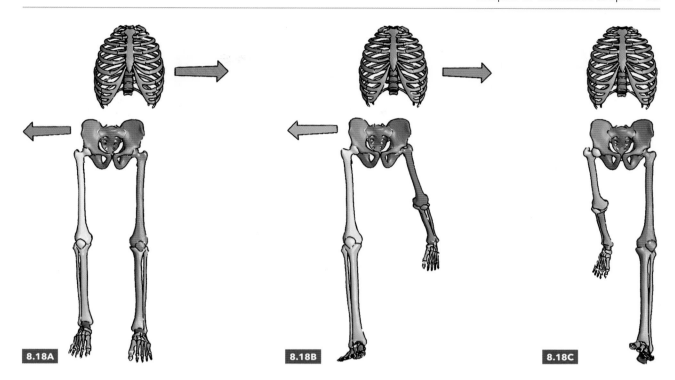

8.18A   8.18B   8.18C

one higher up in the neck, or in the upper thoracic region). Because of the somewhat limited ability of the vertebrae in the lumbar region to shift, in my practice, I usually see that the shift gets accentuated at T11, with the T11 shifting on the T12, due to the change of the angle of the facets, which also affects the relationship between rib 11 and rib 12. It can also be the case that we have no shifting on either side or that the shift we see in a visual assessment stays the same when we look at the person walking, which I have noticed is frequently the case with people with scoliosis.

Opposition in the transverse plane

Just as in the frontal plane, the pelvis and the ribcage also move in opposite directions in relation to each other in the transverse plane. When the pelvis rotates to the left in relation to the ribcage, the ribcage rotates toward the right in relation to the pelvis. This movement illustrates what we usually refer to as contralateral gait. The rotating pelvis follows the anterior tilting of the femur in relation to the pelvis on the trailing leg. So if we have the right leg going forward and the left leg going into an anterior tilt in relation to the pelvis, the pelvis rotates to the left in relation to the ribcage, as the green arrow indicates in figure 8.19. The ribcage follows the swinging arm on the left side

and rotates toward the right in relation to the pelvis, represented by the red arrow turning toward the right.

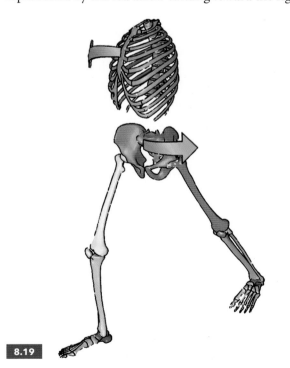

8.19

If we want to make things a bit more complex, we can relate the movement of the ribcage and the pelvis to something else in the body. If we, for example, relate the movements of the ribcage and the pelvis to the right femur, the ribcage sometimes moves more in relation to the

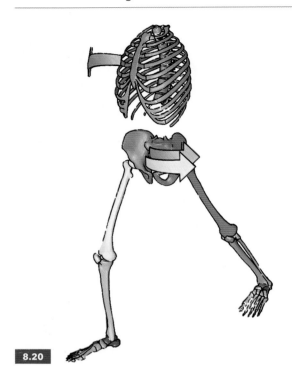

8.20

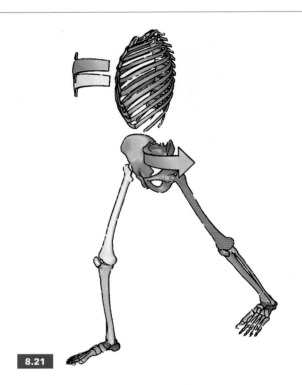

8.21

femur, or the pelvis moves more in relation to the femur. If we look at figure 8.20, the pelvis is moving in relation to the right femur, but the ribcage is not. The yellow arrow indicates that the pelvis is rotating toward the left in relation to the femur. In figure 8.21 we have the opposite where the pelvis is not moving very much in relation to the femur, but the ribcage is. Here the yellow arrow represents the right rotating ribcage in relation to the femur. One thing to remember though is that the relationship between the ribcage and pelvis can still be the same in all these pictures, as the red and green arrows show; what is different is the relationship toward something else, in this case, the right femur. In a later chapter, we will look at all the movements in the transverse plane, and connect the movements in the feet all the way up to the ribcage.

The shifting between the ribcage and the pelvis also involves a strong rotational element. It is easier to keep a shift between the ribcage and the pelvis if the ribcage rotates away from the trailing leg. Stand up and do a shift of the ribcage in relation to the pelvis like that shown in figure 8.18A. Then rotate your ribcage toward the right in relation to the pelvis, and then rotate toward the left in relation to the pelvis. What you will notice is probably that the shift gets accentuated when you rotate toward the right and that it disappears more into a tilting relationship between the ribcage and the pelvis when you rotate toward the left.

Opposition in the sagittal plane

If we follow our underlying premise of opposition between ribcage and pelvis, if we tilt the pelvis anteriorly, the ribcage should tilt posteriorly (in relation to each other). In the world of structural integration when you talk about a posterior tilt of the ribcage it usually also involves a posterior shift of the ribcage.[2] This is not what is meant here. When we talk about the posterior tilt in this context, it does not involve a posterior shift; the ribcage stays centered over the pelvis, or if anything it shifts anteriorly in relation to the pelvis.[3] If the pelvis tilts anteriorly and the ribcage posteriorly without a posterior shift of the ribcage, this means that we get a full posterior bending in the whole spine (extension), which opens the front of the body (figure 8.22). If we are in this fully posteriorly bended position, we can then posteriorly tilt the pelvis in relation to the ribcage with the swinging leg and anteriorly tilt the ribcage in relation to the pelvis, from a posteriorly tilted position (figure 8.23).[3]

When the pelvis goes into anterior tilt, we should see a lift in the ribcage ventrally (posterior tilt of the ribcage). When we see somebody walking this is usually not the case. The posterior tilt and the lift of the ribcage are dependent on the pelvis traveling from posterior to anterior (and the movement of the

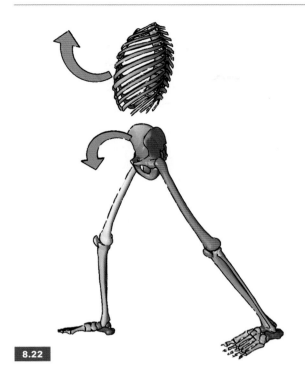

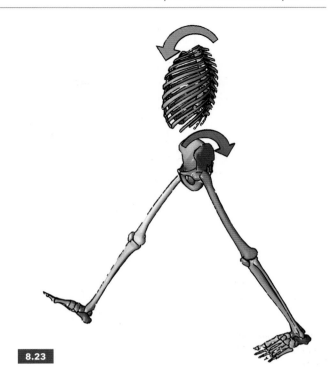

8.22

8.23

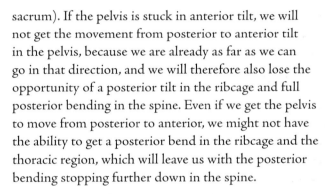

sacrum). If the pelvis is stuck in anterior tilt, we will not get the movement from posterior to anterior tilt in the pelvis, because we are already as far as we can go in that direction, and we will therefore also lose the opportunity of a posterior tilt in the ribcage and full posterior bending in the spine. Even if we get the pelvis to move from posterior to anterior, we might not have the ability to get a posterior bend in the ribcage and the thoracic region, which will leave us with the posterior bending stopping further down in the spine.

Another alternative can be that we are stuck in a position of posterior tilt of the pelvis. Because we are stuck in this position, we will not get any anterior tilting of the pelvis, and therefore also no proper posterior bending throughout the spine. We will not get any movement of posterior tilting of the pelvis either as this requires the pelvis to be in a more anterior position first. If we have proper opposition between the ribcage and pelvis, we should see a lift in front of the spine in the thoracic region when the pelvis is anterior tilting in relation to the ribcage.[4] The lumbar vertebrae then anteriorly shift in relation to the pelvis and the ribcage (which is enhanced by the movement of the sacrum),[5] this then reverses when the pelvis goes into posterior tilting, and we should then instead see a lift in front of the lumbar vertebrae. In a later chapter, we will describe

how these movements relate to what happens in the feet and the legs.

Opposition in the feet

To understand the overall movement in the feet we are going to divide the foot into forefoot and rear foot. We are going to look at the rear foot and the forefoot as if they are one unit, with no movement inside the rear foot and the forefoot. There is obviously a lot of movement going on within the rear foot and the forefoot, but we make this division so that it is easier to get an overview and understanding of the overall global movements in the feet.

If we look at how this divide is made, the rear foot is the talus and the calcaneus and the forefoot is everything in front of that (figure 8.24). To fully understand what happens in the feet we should know exactly how every bone moves in relation to every other bone. This would be extra important if we want to have a clear intent and create lasting effects in our interventions. But a prerequisite for looking at every bone in detail would be to have an overall sense of the global movement in the feet, which we aim for here; we therefore make things simpler and divide the foot into rear foot and forefoot.

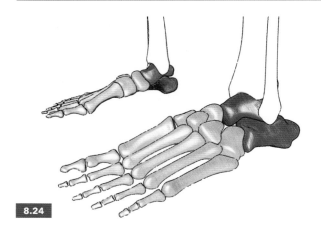

8.24

We will explore two movements in the feet, pronation, and resupination. Pronation and resupination in this context are explained as dynamic events that take place in the whole foot. These movements will be explained as an opposition between the rear foot and forefoot.

Pronation

The first thing to remember is that there is a big difference between the movement of something and being stuck at the end range of that movement. The notion that pronation is something bad or something to be avoided comes up quite often during the interaction with my clients and with the public at large. However, pronation is a key movement in our bodies and what is "bad" is usually being stuck at the end range of that movement. The full motion of pronation happens when the whole of the foot touches the ground (even though it starts a bit earlier). If we examine every plane by itself we will get the following.

Frontal plane

In the frontal plane, the rear foot tilts medially in relation to the forefoot, as in figure 8.25A (the foot on the left is just for reference). The forefoot stays still in relation to the floor, but because of the movement of medial tilt of the rear foot, the forefoot actually tilts laterally in relation to the rear foot. If we were to put the rear foot vertically in relation to the tibia but keep the same

relationship between the forefoot and the rear foot we would see that the forefoot tilts laterally (figure 8.25B). This is a test you could do when somebody is lying on the table: if you put the calcaneus straight in relation to the tibia and look which way the forefoot tilts in the frontal plane it will give you an indication of how the foot behaves in gravity. If the forefoot tilts laterally, as it does in figure 8.25B, it indicates the foot is stuck on the pronation side of things and has

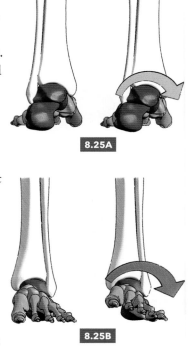

8.25A

8.25B

a hard time with resupination and medial tilt in the forefoot (explained later). If we want a movement of medial tilt of the rear foot in relation to the forefoot when the foot hits the ground, we must first have a lateral tilt or a neutral rear foot before the foot hits the ground. If the rear foot is already very much medially tilted it cannot travel any further, and we cannot get any movement of the rear foot going from lateral to medial tilt.

Transverse plane

In the transverse plane, the rear foot rotates internally in relation to the forefoot (figure 8.26A). When the calcaneus tilts medially it makes it easier for the talus to rotate internally in relation to the calcaneus; this opens up the foot and pushes the navicular laterally in relation to the talus, which makes it easier for the whole forefoot to externally rotate in relation to the rear foot (figure 8.26B). It also unlocks the cuboid, which helps to rotate the forefoot in relation to the rear foot externally. If the foot already is pronated with the forefoot externally

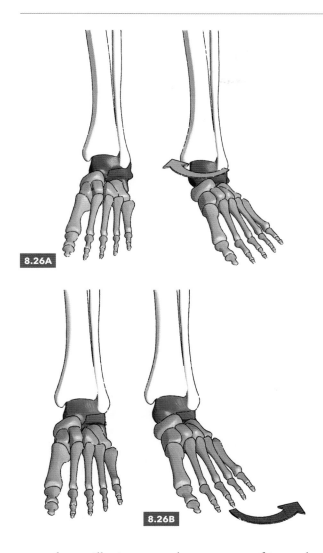

8.26A

8.26B

rotated, we will miss out on the movement of internal rotation of the rear foot in relation to the forefoot.

because the floor stops the forefoot. This makes the foot longer from the calcaneus to the toes, and this is one of the reasons why we see a drop in the medial arch when we pronate. Once again, if the foot is already totally flat in the sagittal plane, the rear foot cannot anteriorly tilt because it is already so far gone in that end range.

Resupination

It is called re(again)-supination because in gait the foot is usually supinated before it pronates, i.e. the foot is supinated before it touches the ground, and after the foot hits the ground and pronates it supinates again. As we said before, being stuck at end range of a movement is not the same as the movement itself. The movement of resupination is dependent on the movement of pronation. If the foot is stuck in supination and does not go into pronation, we are going to have a hard time with the movement of resupination, because we are already in a supinated position. The same goes for being stuck in pronation; resupination is a response to the movement of pronation, and if we do not have any movement of pronation, it is going to be harder with resupination. In resupination, the rear foot posteriorly tilts, laterally tilts and externally rotates in relation to the forefoot. These movements happen together, but if we break it down to every plane of motion, we will see the following.

Sagittal plane

In the sagittal plane, the rear foot tilts anteriorly in relation to the forefoot (figure 8.27). It does so

Frontal plane

In the frontal plane, the rear foot tilts laterally in relation to the forefoot (figure 8.28). The lateral tilt

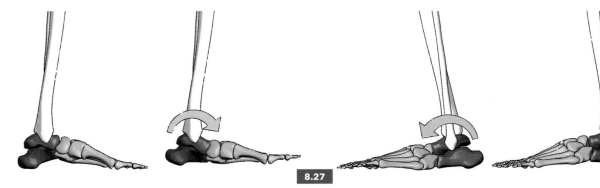

8.27

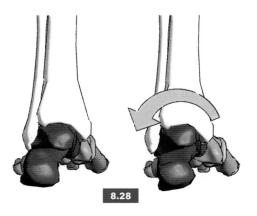

8.28

starts from a medially tilted position, the endpoint of the medial tilt in pronation. The lateral tilting is crucial for making the foot stable so it can propel us forward. When the rear foot laterally tilts the forefoot stays in contact with the floor and we therefore get a medial tilt of the forefoot in relation to the rear foot. If the rear foot always stays medially tilted as in pronation, the tibia tends to follow and we tend to get stuck in an x leg pattern, with the knees medially shifted. A crucial strategy for working with x leg would, therefore, need to involve some form of intervention that takes into account the lateral tilting of the rear foot in relation to the forefoot.

Transverse plane

In the transverse plane, the rear foot externally rotates in relation to the forefoot (figure 8.29). In resupination the talus externally rotates in relation to the calcaneus, which makes it easier for the forefoot to rotate in relation to the rear foot internally. A lot of this rotation happens between the talus and the navicular, and somewhat between the calcaneus and the cuboid.

If you do a visual assessment and see that the person in front of you has their feet in a position where the forefeet are externally rotated in relation to the rear feet, it is probably also the case that they lack this movement of external rotation of the rear foot in relation to their forefoot. If this movement does not occur, we usually see some "compensations" in the transverse plane higher

up in the body. One sign that we lack this movement can be that we stand with the feet turned out as in figure 8.30A, and if we were to turn the forefeet so they are directed straight forward it feels like we are standing with the femurs really internally rotated, as in figure 8.30B. One reason for this is the lack of external rotation of the rear foot in relation to the forefoot (and external rotation of the talus in relation to the calcaneus). The tibia follows the talus and the rear foot, hence the internal rotation of the whole leg if we stand with the forefoot pointed straight forward (red arrows). This can be one reason why we prefer to stand with the feet turned out.

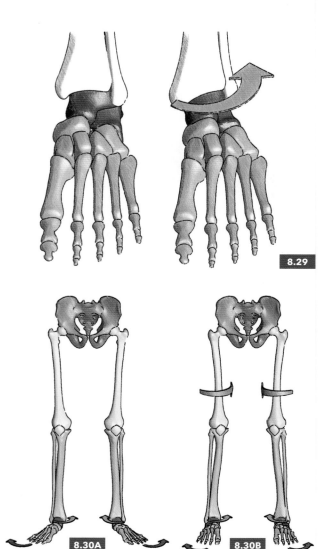

8.29

8.30A   8.30B

Sagittal plane

In the sagittal plane, the rear foot posteriorly tilts in relation to the forefoot, or the forefoot anteriorly tilts in relation to the rear foot (figure 8.31). This makes the foot shorten up again, and the medial longitudinal arch lifts. A big part of this is the anterior tilt of the first metatarsal in relation to the rear foot. When the first metatarsal anteriorly tilts, the navicular drops and anteriorly tilts in relation to the talus.[6] In figure 8.32 this is illustrated by the orange arrows under the navicular in the foot on the left-hand side. This can be a bit confusing because at the same time the navicular superiorly shifts in relation to the ground or the tibia, as in figure 8.32 (represented by the blue arrow in the foot on the left-hand side). When the foot pronates the opposite happens; the navicular shifts superior and posterior tilts in the sagittal plane at the same time as the first metatarsal posterior tilts in relation to the rear foot.

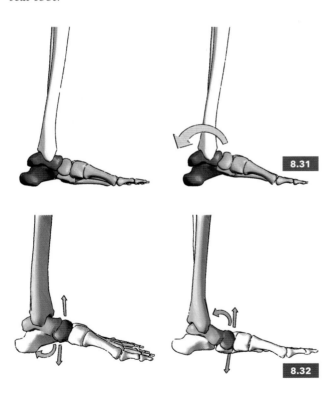

8.31

8.32

When people talk about a navicular drop, they usually have the floor as a reference point. But if you try to lift the navicular with the talus stuck in anterior tilt (and medial tilt and internal rotation) you are probably going to make things worse, as this makes it easier for the first metatarsal to posteriorly tilt (dorsiflex) in relation to the rear foot, making you end up with even more of a "navicular drop." This is, on the other hand, a good strategy if you want a foot stuck in supination to start to pronate. If you are working with a foot that is stuck in pronation, you should instead try to inferior shift, anteriorly tilt, medially tilt/shift and internally rotate the navicular in relation to the talus and the rear foot (and posterior tilt the rear foot in relation to the tibia). This would make it easier for the first metatarsal to anteriorly tilt. The resupination movements of the lateral tilt (figure 8.28) and the external rotation (figure 8.29) of the rear foot also make it easier for the first metatarsal to anteriorly tilt.

Posterior tilt of the big toe

The anterior tilt of the first metatarsal in relation to the rear foot and the posterior tilt (dorsiflexion) of the big toe are intimately connected. In the later stages of resupination, only the forefoot is on the ground, the forefoot medially tilts and the rear foot laterally tilts in relation to each other; this makes it easier for the big toe to posterior tilt. The next time you have somebody's foot available, laterally tilt the forefoot in relation to the rear foot and ask them to posterior tilt the big toe, and then medially tilt their forefoot and ask them to posterior tilt the big toe. Probably you are going to find that it is much harder to posterior tilt the big toe with the forefoot laterally tilted. When you laterally tilt the forefoot you also unavoidably posterior tilt the first metatarsal in relation to the rear foot; this is what makes it harder to posterior tilt the big toe. It is the same thing if you stand up and put one foot in a supinated position and lift the big toe, as in figure 8.33, and then put the foot in a pronated position, as in figure 8.34 and then lift the big toe. What you will probably find is that it is harder to lift the big toe when you are in a pronated position.

In this context it is better to think about posterior tilt of the big toe as two movements: posterior tilt of the

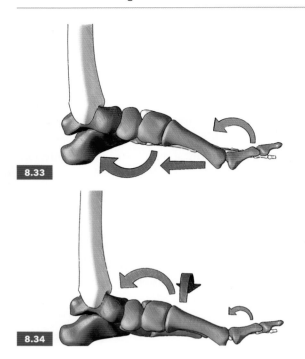

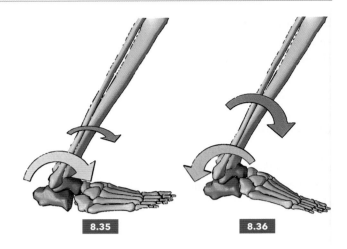

to resupinate (the orange arrows demonstrate the relationship between the rear foot and forefoot and the blue arrows the relationship between the rear foot and the talus).

big toe in relation to the first metatarsal, and anterior tilt of the first metatarsal in relation the rear foot. These two movements are connected to resupination in the foot and shorten up the foot and bring the head of the first metatarsal closer to the rear foot, as indicated by the blue arrow behind the head of the first metatarsal in figure 8.33.

Talus and tibia relationship in the sagittal plane

In a foot that is stuck in a pronated position the rear foot is chronically anteriorly tilted in relation to the forefoot. A prerequisite for getting the rear foot to posteriorly tilt in relation to the forefoot can sometimes be to get enough posterior tilt of the rear foot in relation to the tibia. If we have limitations in this movement, the rear foot can be forced to anteriorly tilt in relation to the forefoot instead of posteriorly tilting. In figure 8.35 the rear foot is anteriorly tilting in relation to the forefoot (orange arrow), and the tibia is anteriorly tilting in relation to the talus very little, indicated by the small blue arrow. If we were to have more movement between the talus and the tibia, the rear foot would have a better chance to posteriorly tilt in relation to the forefoot, as in figure 8.36, and it would be easier for the foot

A further compensation if we lack anterior tilt of the tibia in relation to the rear foot can be that the rear foot leaves the ground earlier than it would otherwise do. We do this with a combination of anterior tilt of the rear foot in relation to the forefoot and anterior tilt of the tibia in relation to the femur as in figure 8.37. The anterior tilt of the tibia in relation to the femur makes it easier for the rear foot to lose contact with the ground a bit earlier, so we can move forward with less movement between the rear foot and the tibia. We are going to look more closely at how these movements relate higher up in the body in a later section.

When the rear foot is always anteriorly tilted, as when the foot is stuck in pronation, the tibia and fibula tend to move posteriorly on the talus. Because of the wedge shape of the talus being wider anteriorly[7] than posteriorly, the malleoli needs to separate from each other when the tibia dorsiflexes in relation to the talus.[8] If the rear foot is chronically anteriorly tilted in relation to the tibia and the forefoot this might make it harder for the malleoli to separate because they are never challenged. Resupination in the foot would challenge the movement between the talus, the tibia and the fibula, at the same time as posterior tilt of the rear foot in relation to the tibia would make it easier for us to get the resupinating movements in the rest of the foot.

Talus and tibia relationship in transverse plane

It is my belief that there also is a strong rotational element between the talus and the tibia/fibula in feet that are stuck in pronation. In a foot that is stuck in a pronated position, the talus is chronically internally rotated both in relation to the calcaneus, but also to the forefoot and the tibia, as in figure 8.38. If we focus on what happens between the talus and the tibia, and how it relates to anterior tilt of the tibia in relation to the talus, the internal rotation of the talus in relation to the tibia makes it harder for the lateral side of the inferior articulate surface of the tibia to glide over the talus. In figure 8.39, the foot on the left-hand side is a foot in a "neutral position." The second foot in the middle is the same foot but without the forefoot, and in the foot on the right, the talus is internally rotated in relation to the tibia, or the tibia has laterally rotated in relation to the talus, which is indicated by the orange arrow (the arrow describes a position, not a movement).

In the foot that is stuck in pronation, the medial side of the tibia is in a position where it has shifted forward in relation to the talus, with the lateral side not coming along but stuck posterior in relation to the talus. When we try to anterior tilt the tibia in relation to the talus, the lateral side of the tibia does not want to slide over the talus. The black arrow in figure 8.40 shows where this happens. The lateral border of the trochlea (the superior part of the talus) is also supposedly longer, which suggests that the lateral side of the tibia has a longer way to travel than the medial side.[9] If we try to posterior tilt the whole foot in relation to the tibia with somebody lying down who has a foot that is stuck in pronation, we are probably going to find that it is easier to posterior tilt the foot in relation to the tibia when the foot is in a pronated position. If we instead put the foot in a more supinated position, we are probably going to find that it is much harder to posterior tilt the foot in relation to the tibia. If we want to have proper movement between the talus and the tibia, our interventions should also involve this rotational element between the talus and the tibia. I have found in my clinic that it seems not only beneficial to work with the rotation between the talus and the tibia, but also that working with the relation the talus has to the navicular and the calcaneus can sometimes greatly enhance the movement between

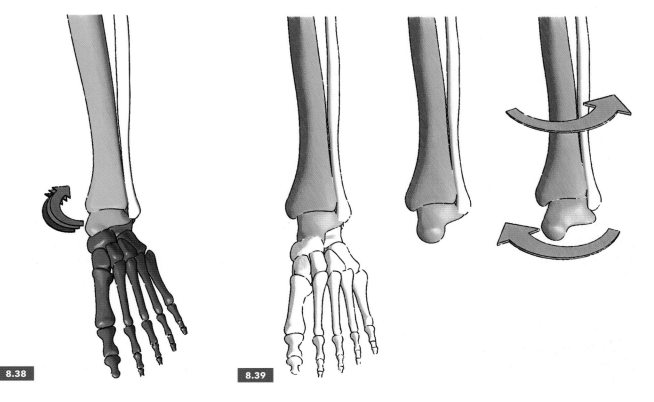

8.38

8.39

the talus and the tibia. All this indicates that resupination in the foot and proper movement between the talus and the tibia are interlinked.

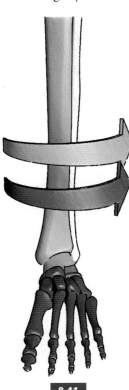

8.40

Another common phenomenon that I have noticed is that if you get a foot that is stuck in pronation to start to resupinate and the foot ends up in a more supinated position, the rotation between the talus and the tibia becomes more apparent. Looking at figure 8.38 again, the tibia is slightly internally rotated in relation to the forefoot. If we were to resupinate the foot so that the talus ends up in a position where it is more externally rotated in relation to the calcaneus and the navicular, the tibia would end up in a position where it is more externally rotated in relation to the forefoot. In figure 8.41 the talus has rotated in relation to the forefoot and the calcaneus into a more supinated position, this results in the tibia being more externally rotated in relation to the forefoot (blue arrow). This can make the rotation between the talus and the tibia appear more "visible" (because the internal rotation of the talus does not hide it). The relationship between the talus and the tibia is still the same, though; the yellow arrow shows the position of the tibia in relation to the talus (not a movement).

8.41

Counterintuitively, sometimes this even affects the rotation between the tibia and the femur, leaving us with more external rotation of the tibia in relation to the femur. If the femurs do not allow for more external rotation in relation to the pelvis, the rotation can instead end up more between the tibia and the femur. To counteract this, we can work with externally

rotating the femur in relation to the pelvis and externally rotate the talus in relation to the tibia, or in other words, internally rotate the tibia in relation to both the femur and the talus.

Putting it all together

We are now going to look at how the movements of the feet relate to the rest of the body, or how movement in the rest of the body relates to the feet. Primarily we are going to look at how the feet relate to movement between the femur and the pelvis, and opposition between the ribcage and pelvis.

Resupination in the transverse plane

We will begin by looking at how different movements relate in the transverse plane. The resupination of the foot, i.e. external rotation of the rear foot in relation to the forefoot, has a direct relation to the external rotation of the femur in relation to the pelvis. If the foot resupinates properly we get more external rotation of the femur in relation to the pelvis, as in figure 8.42. This also means that the pelvis rotates toward the left in relation to the femur (shown by the yellow arrow). If we do not have any resupination in the foot, we will get less external rotation of the femur. In figure 8.43, where the foot is in a pronated position, we instead get an internal rotation of the femur in relation to the pelvis, or right rotation of the femur in relation to the femur.

8.42

If we look a bit more closely at what happens between the rear foot and the tibia, we can see in figure 8.44 that the rear foot externally rotates in relation to both the forefoot and the tibia, as the orange and light blue arrows show. This is the same as saying that the tibia is internally rotating in relation to the rear foot as is shown here by the blue arrow over the tibia. The tibia also internally rotates in relation to the femur as the yellow arrow shows; this is the same movement as the blue arrow over the femur demonstrates. The femur is externally rotating in relation to both the tibia and the pelvis. If we want to put it simply, if the forefoot, tibia, and pelvis "stay still," the rear foot and the femur externally rotate in relation to all of them.

One frequent pattern that I find in my clinic is what can be called x legs with pronated feet. This is the same pattern as the one in figure 8.30A but a bit more nuanced. If you look at figure 8.45 you can see that the forefoot is externally rotated in relation to the rear foot, the rear foot is internally rotated in relation to the forefoot, the tibia is externally rotated in relation both to the rear foot and the femur, and the femur is internally rotated in relation to both the tibia and the pelvis (this describes a static position). It is called x legs because of the medially shifted knees; the knee shifts medially partly because of the rotation between the tibia and the femur. The forefeet are turned out, and if the person were to put the forefoot straight ahead, the femur would get even more internally rotated (as we described in figure 8.30B). The position in this pattern describes precisely the opposite to the movements we described above when we described the pattern of resupination up the body in the transverse plane (figure 8.44). It would therefore

not be any surprise that this is the movement that this pattern probably lacks;[10] to really know though, we would have to do a proper movement assessment and see how the body moves.

If we look at how this pattern behaves in the transverse plane in gait, we will get something opposite to what we described before. If we look at figure 8.46, the rear foot is internally rotating in relation to both the forefoot and the tibia, the tibia is externally rotating in relation to the femur and the rear foot, and the femur is internally rotating in relation to both the tibia and the pelvis. As long as we have a lack of resupination (external rotation of the rear foot) we are also going to have a hard time with external rotation of the femur in relation to the pelvis. Because when the rear foot internally rotates it pushes the whole leg into an internal rotation in relation to the pelvis, hindering us from getting any external rotation of the femur in relation to the pelvis. A solution to not having the femurs totally internally rotating can be to turn the feet out when we walk; this will leave us feeling a bit less strained and will make it a bit easier for us to move forward. However, this usually means that we externally rotate the tibia in relation to the femur instead of externally rotating the femur in relation to the tibia; this will also make it much harder for us to resupinate. This movement pattern usually corresponds to the aforementioned static position.

Many times the lack of external rotation of the femur can also be a hindering factor in getting the foot to resupinate. When working in my clinic, it has been a common theme that you work with resupinating the feet, but nothing happens until you get the femurs to rotate externally in relation to the pelvis. Working with external rotation of the femur, usually also makes it easier for the person to move the pelvis from anterior to posterior tilt. One reason for this is that external rotation of the femurs in relation to the pelvis pushes the inferior part of the pelvis anterior in relation to the head of the femurs (or in relation to space itself or the feet), making it easier for us to posterior tilt the pelvis.

If we continue our way up the body and look at the relationship between the femur, the pelvis and the ribcage, we can see that in figure 8.47, the pelvis rotates to the left in relation to the femur but to the right in relation to the ribcage. This is a way to describe maximum length, "twist," or load up in the transverse plane, between the femur and the ribcage. The external rotation of the femur is also one aspect that not only drives the differentiation between the femur and the pelvis but also between the pelvis and the ribcage and acts as a counter movement for the rotating ribcage. If the femur is not externally rotating in relation to the pelvis as in figure 8.48, the pelvis is not rotating as much to the right in relation to the feet (or the way we are travelling), which means that we will also get less separation between the ribcage and pelvis. That is if the ribcage

8.47

8.48

does not move "even more" to the left because the ribcage is not hindered by the right rotating pelvis and the externally rotating femur. This is the same phenomenon that we described in figures 8.20 and 8.21, where the ribcage rotated to a higher or lesser degree in relation to the right femur. But we can now see that this is connected to the amount of resupination we have in the foot. If we don't have any resupination in the foot we will get less external rotation of the femur in relation to the pelvis, as in figure 8.48, and we will therefore not have anything that is moving in the "opposite" direction as the ribcage, and therefore also less separation between the femur and the ribcage in the transverse plane.

To explain this further, in figure 8.49, the pelvis and the ribcage rotate toward the left in relation to the left femur (pink arrows). The ribcage moves a bit more represented by the bigger pink arrow in front of the ribcage. They both rotate toward the left in relation to the left femur due to the lack of resupination, as the rear foot does not externally rotate in relation to the forefoot, and also because of the lack of external rotation of the femur in relation to the pelvis that comes with this. Because of this lack, we do not have anything rotating the pelvis toward the right in relation to left femur; the pelvis then almost rotates the "opposite way", following the ribcage toward the left in relation to the left femur, indicated by the pink arrows. As you can see, the resupination or the lack of resupination would also affect the relationship between the pelvis and the left femur. The lack of resupination would not only create less external rotation of the right femur in relation to the pelvis but also less external rotation of the left femur in relation to the pelvis. If we want to put it in simpler terms, we can say the resupination in the right

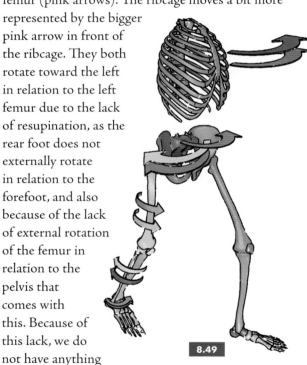

8.49

foot opens up the space between the head of the right femur and the left femur in the transverse plane. The lack of resupination instead closes the space in the transverse plane between the head of the femurs. In figure 8.50 you can see that the pelvis rotates away from left femur when the foot is resupinating.

The fact that the lack of resupination not only affects the relationship between the femur and the pelvis

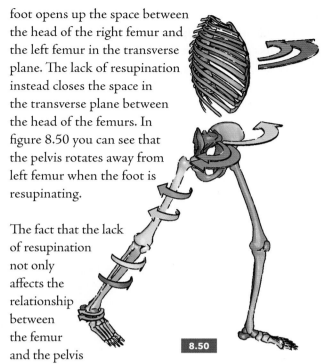

8.50

on the same side but also affects the other side, is one of the reasons it can be so confusing to understand why the pelvis is rotated toward the left or the right (in relation to the feet or the ribcage for example) in a visual assessment and what movement or lack of movement is connected to that rotated pelvis. But one simple guideline is that usually, the pelvis rotates toward the side where we have more movement of pronation and especially resupination, and that the pelvis rotates away from a foot that is stuck in pronation, where we have no movement of pronation and resupination. The lack of pronation and resupination is also very frequently accompanied with a movement of pronation when we actually should have a movement of resupination, as in figures 8.46, 8.48, 8.49; the pelvis usually also rotates away from this side of "false" pronation.[11] If we want a more detailed understanding of what movements or lack of movements are related to the rotated pelvis in any given case, we need to do a proper movement and gait assessment and see what is actually happening (strategy 3).[12] Of course, all rotations do not need to be only movement-related, and we can have structural asymmetries in the skeleton (or other factors). A detailed gait assessment can give us indications of this.

Resupination in the frontal plane

Next, we are going to look at relationships in the frontal plane. When the foot resupinates and the rear foot

laterally tilts in relation to the forefoot the pelvis tilts in the frontal plane toward the same side. If we look at figure 8.51, we can see that the rear foot laterally tilts in relation to the forefoot and that the pelvis is tilting toward the right in relation to both the left and the right femur and the ribcage. This means that we get a left tilt of the right femur in relation to the pelvis. The resupination in the foot, the tilting of the pelvis, and the tilting of the ribcage creates length in the frontal plane and means that we get a load up in the frontal plane. If we did not have the resupination in the foot, we would miss out on this length in the frontal plane. If the rear foot is medially tilting instead of laterally tilting, we do not get the same amount of medial tilt of the right femur in relation to the pelvis, or tilt of the pelvis in relation to the femur. In figure 8.52, we get less of a tilting of the pelvis in relation to the right femur (small yellow arrow) but more so in relation to the left femur and the ribcage. The resupination and the medial tilt of the right femur in relation to the pelvis are really what controls and opposes the tilting of the pelvis in relation to the left femur and the ribcage.

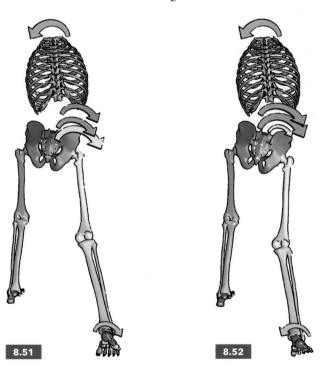

8.51 **8.52**

If we have anterior tilt of the tibia in relation to the femur, this also opens up the possibility for the knee to medially shift. If you stand up and try to medially shift the right knee when you are in a position like the one in figure 8.51, with tibia as posteriorly tilted as possible in relation to the femur (knee extended), you are probably going to find it really hard. If you instead bend the knee and try to medially shift the knee you are probably going to find that you can go very far, especially if you lift the heel a bit. When the knee is bent it really can start to move more in different directions; the external rotation of the tibia in relation to the femur, as described in the last section, is also involved when we shift the knee medially. If somebody does not resupinate, it is therefore not very strange if we find them in an x leg position, with the knee medially shifted, as described earlier.

Another static position that is very common is that the pelvis shifts anteriorly in relation to the feet. The lack of resupination can be a reason why this is the case. If we have the rear foot medially tilting (8.52) instead of laterally tilting (8.51), this means that the forefoot is also laterally tilting in relation to the rear foot. If we do not have any support from the first metatarsal when standing up in a static position, we do not have anything "pushing" the pelvis posterior. The lateral tilting of the forefoot creates less of this support, which means that we are more prone to end up in this position, as in figure 8.53. All the transverse relationships described in figure 8.45 and the correlating movements (or lack of movements) in the transverse plane described, create less support from the first metatarsal as well, which can create more of an anterior shift.

8.53

Resupination in the sagittal plane

We are now going to look at how things relate in the sagittal plane. One essential thing to look at when we do a gait assessment is anteriorly tilting of the femurs in relation to the pelvis (extension at the hip joint). When the foot resupinates it makes the foot more stable, which propels us forward in a better way, which also drives the leg into more anteriorly tilting in relation to the pelvis. It can be a little difficult to see because we usually see the person walking away and toward us. However, one thing to look at is how much the back of the femur/ hamstrings come back when the client walks away from us, or how much of the femur disappears or how much "the hip" opens up when the client walks toward you. A good thing to know is that usually anterior tilting of the femur in relation to the pelvis comes with resupination in the foot and external rotation of the femur in relation to the pelvis, and more rotation of the pelvis toward that same side as we described in the earlier section.

When the foot resupinates, the rear foot posteriorly tilts in relation to the forefoot. Once again this movement is "opposing" what happens further up the body, which means that we get more length and load up in the sagittal plane. This is illustrated in figure 8.54, where the rear foot is posteriorly tilting in relation to the forefoot (orange arrow) and the pelvis is posterior tilting in relation to the femur (yellow arrow); this is the same as saying that the femur is moving into anterior tilting in relation to the pelvis. The pelvis is anteriorly tilting in relation to the ribcage, and the ribcage is posteriorly tilting in relation to the pelvis. If we want to relate it to something else in the body, we can see that the pelvis is also anteriorly tilting in relation to the left femur and that the ribcage is also posteriorly tilting in relation to the left femur. When the femur anterior tilts in relation to the pelvis, the pelvis anteriorly tilts and the ribcage posteriorly tilts in relation to each other; this creates maximum "length" in the sagittal plane. If, for example, we would have less anterior tilting of the femur, or less anterior tilting of the pelvis, or less posterior tilting of the ribcage, we would have less "length" in the sagittal plane. This is the moment when we get the lift in front of the thoracic spine as I mentioned when we talked about the opposition between the ribcage and the pelvis. If we look at figure 8.55, we can see a lift in front of the thoracic spine indicated here by the orange arrow (in relation to space itself or the feet). At the same time as

8.54

this happens, the lumbar vertebrae shift anteriorly in relation to both the ribcage and the pelvis. Another way to explain this, as I have done before, is that we get a posterior bending throughout the whole spine.

If the rear foot is not posteriorly tilting but instead anteriorly tilting in relation to the forefoot, we lose the "opposing" movement, and we end up with less length and load up in the sagittal plane. This is the same phenomenon that we described in figure 8.37, where the tibia anterior tilts in relation to the femur. This also usually means that we get less anterior tilting of the femur in relation to the pelvis, and no lift in front of the thoracic spine. The ribcage stays still, or posteriorly tilts very little in relation to the left femur. The pelvis posteriorly tilts very little in relation to the right femur, and anteriorly tilts a lot in relation to the left femur. In figure 8.56, we can see that we have a small yellow arrow, showing that the pelvis is posteriorly tilting very little in relation to the femur, and the bigger pink arrow showing that the pelvis is anterior tilting a great deal in relation to the left femur. Instead of getting the posterior bending throughout the whole spine, we instead get accentuated posterior bending in the lumbar vertebrae because of the increased anterior tilting of the pelvis and the lack of anterior tilting of the right femur and the lack of lift in the thoracic spine (small pink arrow). This means that we get the anterior shift of the lumbar vertebrae without the posterior bending in the thoracic vertebrae. A person like this walks forward with a lot of anterior movement

8.55

8.56

of the pelvis, but with less movement between the pelvis and the right femur, and less movement of posterior tilt of the ribcage in relation to the left femur (posterior bending in the thoracic spine). If we compare figures 8.54 and 8.56, we can see that there is a lot more "length" and "lift" in figure 8.54 in the sagittal plane.

When we look at somebody walking, we can see how long they can keep the tibia straight in relation to the femur; if the tibia stars to anterior tilt in relation to the femur very early and we have no anterior tilting of the femur in relation to the pelvis, this is an indicator that we do not have any resupination in the foot. Alternatively, we do not have any resupination in the foot because we do not have any anterior tilting of the femur in relation to the pelvis. Again, if we do some movement assessment of the person, this can help us to get a clearer picture of where we have the most limitations.

We are now going to look at what happens when the pelvis is posteriorly tilting, and how opposition between the pelvis and ribcage relate to this. When we swing the leg forward the pelvis posteriorly tilts both in relation to the ribcage and the femur, as in figure 8.57. At the same time the foot resupinates with movement between the tibia and the talus, as shown here by the blue arrow (this is the same as I described in figure 8.36). When this happens we get a lift in front of the lumbar region, as in figure 8.58; we then also get a posterior shift of the lumbar region in relation to the pelvis and the ribcage.

If we have no resupination in the foot, and no anterior tilting of the femur in relation to the pelvis, we will usually also have trouble with posteriorly tilting the pelvis in relation to the ribcage. If we look at figure 8.59, we can see that we lack resupination in the foot, anterior tilting of the femur in relation to the pelvis, and posterior tilt of the pelvis in relation to the ribcage. In figure 8.37, I described that the heel might lift from the ground prematurely because of the lack of movement between the tibia and talus and that this might also affect our ability to resupinate the foot. We can now clearly see that this also affects the amount of anterior tilt we get in the femur in relation to the pelvis. With resupination in the foot and proper movement between the tibia and the talus, we do not need to bend the knee, and hence it is therefore also easier to get an anterior tilt of the femur in relation to the pelvis (if the body allows it). We also miss out on the lift as described in figure 8.58; if we compare

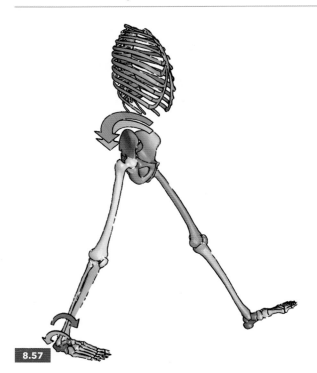

8.57

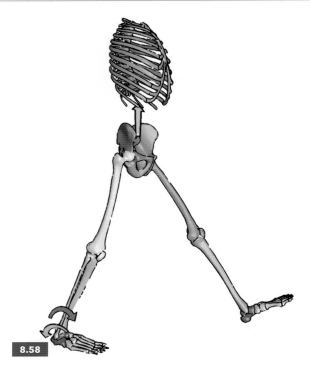

8.58

figure 8.59 to 8.58 we can clearly see that there is more lift and length in figure 8.58. Another aspect is also that if we do not have any posterior tilting of the pelvis, we are set up in a less optimal position for anteriorly tilting the pelvis. If we are already very anteriorly positioned when the pelvis is anteriorly tilting it does not have very far to travel, which will affect our ability to posteriorly bend throughout the whole spine, as I discussed earlier.

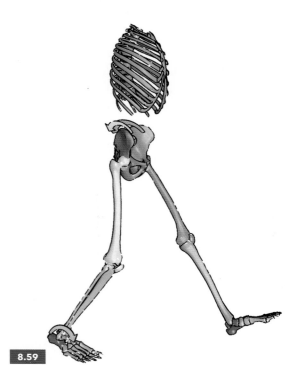

8.59

The fact that we are not bending the knee if we get proper resupination and anterior tilting of the femur also means that the leg gets longer or "higher"; this height generally translates into more of a shift of the pelvis in the frontal plane in relation to the foot, as well as tilt of the pelvis in relation to the femur. In figure 8.60, the pelvis shifts in relation to the foot in the right leg and the pelvis tilts to the right in relation to the femur, because of the shift (blue and yellow arrows). When the pelvis shifts, the right leg gets "shorter," or put in other terms, the distance between the head of the femur and the ground diminishes

8.60

in height. If we do not have any resupination in the foot, as in figure 8.59, we miss out on this movement in the frontal plane. Figure 8.60 is the same picture as figure 8.57 but seen from behind; the front leg is in the air driving the pelvis posteriorly.

Another phenomenon that is also related to the lack of proper movement between the tibia and the talus when the foot is resupinating, as I described in figure 8.36, is anterior tilt of the femur in relation to the tibia. When we swing the back leg forward, it starts to drive the pelvis into posterior tilting, and at the same time, the foot that is on the ground hopefully starts to resupinate. So if we have proper movement between the tibia and the talus we end up in a position like that in figure 8.57. If we do not have resupination in the foot and proper movement between the talus and the tibia, we might instead end up like figure 8.59. There is a moment before we end up like in figure 8.59 when we still have the whole foot on the ground, where restriction between the tibia and the talus can make the femur anteriorly tilt in relation to the tibia instead of the tibia anteriorly tilting in relation to the talus. If we look at figure 8.61, the femur is anteriorly tilting in relation to the tibia and the rear foot is anteriorly tilting in relation to the forefoot. Again if we have a proper anterior tilt of the tibia in relation to the talus, we do not need the femur to anteriorly tilt in relation to the tibia or the rear foot to anteriorly tilt in relation to the forefoot, for us to move forward. In my experience, it is common to see this pattern with the lack of anterior tilting of the femur in relation to the pelvis, because the pelvis anteriorly tilts before we get a chance to get some separation between the femur and the pelvis. It is also common to see this pattern with feet where the rear foot is rigid and held in a supinated position in relation to the tibia at the same time as the forefoot is very mobile and easily laterally tilts in relation to the rear foot.

8.61

Pronation in the transverse plane

In figure 8.46 we described how the tibia externally rotates in relation to the femur and that this movement is something that we do not want when the leg is starting to get behind us. There is a time though when we do want this movement, and that is when the foot is pronating. If we look at figure 8.62, the tibia is externally rotating in relation to the femur at the same time as the foot is pronating, i.e. the rear foot is internally rotating in relation to the forefoot. The pelvis rotates away from the femur when the foot is pronating, which is totally dependent on the resupinating foot in the back leg. If we do not have resupination in the foot in the back leg, we will instead get a pelvis that is rotating toward the left in relation to the femur (internal rotation of the femur in relation to the pelvis), as I described in figures 8.49 and 8.50. The resupination on the back leg and the pronation on the front leg means that both rear feet will rotate toward the right in relation to their forefeet (figure 8.63). It is therefore natural that the pelvis wants to rotate toward the right (in relation to the forefeet or the ribcage). If we instead have pronation on the back leg when we are supposed to have resupination, we close the available "length" in the transverse plane, with both rear feet rotating internally (figure 8.64), making both femurs want to rotate medially in relation to the pelvis.

8.62

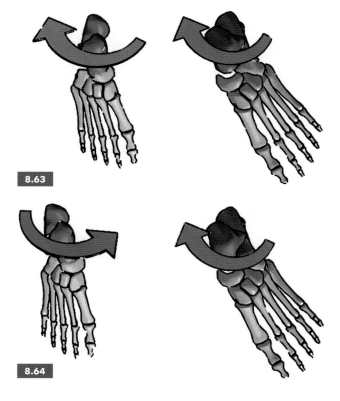

8.63

8.64

Pronation in the frontal plane

As mentioned before, when we bend the knee, it is possible to get more movement between the tibia and the femur. When the foot pronates and the rear foot medially tilts in relation to the forefoot, the tibia follows the rear foot, which makes it possible for some slight medial shift of the knee in relation to the foot and the pelvis. If we look at figure 8.65, the knee medially shifts at the same time as the foot is pronating. The shifting of the knee is very different when we have support from the first metatarsal, which makes it possible for the soft tissue to control the shift. If the foot is already in a pronated position when it hits the ground, the knee can travel a long way medially before we get contact with the first metatarsal, which means that we cannot go any further when the first metatarsal touches the ground, because that would be "too far." Because of this, we lose the opportunity of getting the movement of pronation in the foot. In figure 8.66, the knee has travelled very far medially but the first metatarsal has not yet touched the ground. The orange arrow over the left rear foot describes the position of the medially tilted rear foot in relation to the forefoot, as we do not get any proper movement of pronation between the rear foot and the forefoot until the first metatarsal touches the ground, and in this case the first metatarsal hits the ground very late with the knee already substantially medially shifted,

leaving no room for the movement of pronation to happen. The movement of pronation in the foot and the tilt of the pelvis effectively loads up the whole leg.

Pronation in the sagittal plane

If we look at the sagittal plane, we can see that something similar happens. Both the rear foot and the pelvis anteriorly tilt at the same time as we are bending the knee. The rear foot anteriorly tilts in relation to the forefoot and the tibia anteriorly tilts in relation to the talus; the knee shifts forward in relation to the pelvis and the foot, and the pelvis anteriorly tilts in relation to the femur (figure 8.67). If the rear foot is already anteriorly tilted in relation to the forefoot, and the pelvis is already anteriorly tilted in relation to the ribcage and the pelvis, we cannot go any further, and we lose out on these movements. The anterior tilting of the rear foot and the pelvis are interlinked and this is a way for the body to shock absorb and load up the body. So if we have a foot that is stuck in a supinated position with the rear foot posteriorly tilted

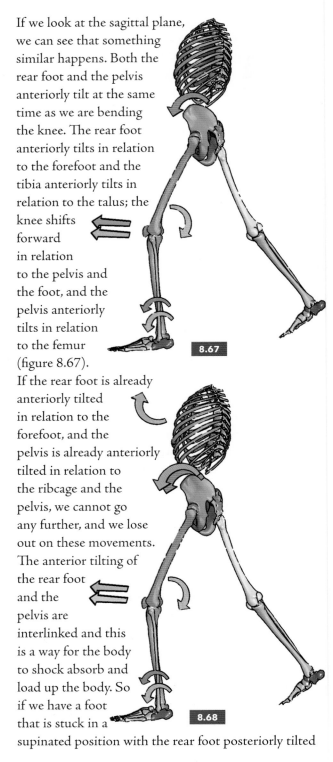

8.65   8.66   8.67   8.68

in relation to the forefoot, we will also miss out on the anterior tilt of the rear foot and its relation to the pelvis. If we do not anteriorly tilt the rear foot and the pelvis we might have to find another way to shock absorb; we perhaps anteriorly shift the knee a bit more or we anteriorly bend the thoracic spine instead of posteriorly bending it, something that is very common if you look at runners. The movements in the upper body are the same as I discussed in figure 8.54. If we look at figure 8.68, the pelvis is anteriorly tilting in relation to the ribcage and the femur, and the ribcage is posteriorly tilting in relation to the pelvis. Again this is a way to describe posterior bending throughout the whole spine, and this is also interlinked to what happens to the rear foot in the pronating foot.

Practical considerations

We are now going to go through some practical guidelines that can help out when we work with people. These are more practical suggestions of how things might relate in the body more than they are exact rules that apply in every case. There can be a lot of unforeseen aspects and complexities in any given body.

If we find somebody with a ribcage that is tilted to the left and rotated to the left in relation to the pelvis in a visual assessment, as in figure 8.69, this is accentuated when the person walks and the ribcage is only tilting and rotating toward the left but not the right in relation to the pelvis. If we do a movement assessment and see that the person has a hard time rotating and tilting the ribcage toward the right in relation to the pelvis, it would be quite natural to do some form of intervention that involves these movements or makes it easier for the person to access these movements. However, if we look at Gait Made Simple, we can see that this lack of tilting and rotating toward the right might also be related to what happens further down. One easy thing to check is if there is a difference in the person's ability to anteriorly tilt the femur in

8.69

relation to the pelvis. If we find that the person has a hard time anteriorly tilting the left femur in relation to the pelvis we know that this limitation could be involved in the inability to tilt and rotate the ribcage toward the right, as these movements are interlinked. The limited anterior tilting can also affect the pelvis rotation toward the left in relation to the ribcage. If we look at figures 8.50, 8.51, 8.54, we can see that these movements are related. If we continue to explore what movements can be related to this, we can see that resupination of the left foot, proper anterior tilting of the tibia in relation the talus in a supinated position, as in figures 8.57 and 8.36, and external rotation of the femur in relation to the pelvis can be some movements that affect our ability to tilt and rotate toward the right.

To take another example, let's say that a person has a ribcage that is shifted toward the right and rotated toward the left when we do a visual assessment, as in figure 8.70. Now, this rarely comes in isolation; there is often also a tilt of the ribcage and elements happening within the ribcage, but for simplicity we leave that for now. In gait, we see that the ribcage shifts toward the right and rotates toward the left in relation to the pelvis, but not the other way. When we do a movement assessment, we see that the person has a hard time shifting the ribcage toward the left and rotating toward the right in relation to the pelvis. The easiest thing would be to work with some form of interventions to make it easier for the person to shift left and rotate right with the ribcage in relation to the pelvis. If we want to look at how it relates to the rest of the body and take a look at the feet, for example, we can see that the left shift and right rotation of the ribcage that we are lacking happens when we are resupinating on the left foot at the same time as we are swinging the right leg forward and posteriorly tilting the pelvis and shifting it toward the right in relation to the ribcage. This is what is described in figures 8.58, 8.57, and 8.60 (except that in these pictures the situation is reversed and we have the right leg behind us). If we do a gait assessment and

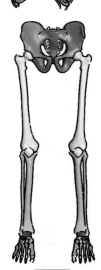

8.70

some movement assessments we can see if the person can access these movements properly and see which movements are missing. Can the person resupinate the left foot, externally rotate, medially tilt, and anteriorly tilt the left femur in relation to the pelvis? Another thing to look at is if the person is pronating on the right leg when we are supposed to resupinate. If we are pronating as in figure 8.59 we do not have anything controlling and opposing the shift of the pelvis toward the left, we get less medial tilt of the femur in relation pelvis, and we therefore might need to work on the resupination on the right side as well.

If we find somebody in a pattern with medially shifted knees (x legs) and pronated and turned out feet, as in figure 8.71, we might want to assess their ability to resupinate and track it up to the pelvis. In figure 8.71, the feet are pronated, and the tibia is externally rotated in relation to both the talus and the femur; the femur is internally rotated in relation to the pelvis, and the pelvis is anteriorly tilted in relation to the femur. The exact opposite movements of this position are found in figure 8.57 (the transverse movements are the same as in figure 8.44). The feet resupinate, the talus externally rotates in relation to the tibia, the tibia anteriorly tilts in relation to the talus, the tibia internally rotates in relation to the femur, the femur externally rotates in relation to the pelvis, and the pelvis is posteriorly tilting in relation to the femur and the ribcage. All these movements are interlinked, and it might be hard to get the pelvis to posteriorly tilt more easily, for example, without working with all these movements.

In figure 8.72 we have another example that is quite common. In this picture, the pelvis is in a position where it is posteriorly tilted in relation to the ribcage as well as the femur. The ribcage is anteriorly tilted in relation to the pelvis. This is a way to describe that we

8.71

have an anterior bend in the whole spine (maybe not in the cervical region). If we look at the relationship between the ribcage and the pelvis, we could start by assessing the person's ability to anteriorly tilt the pelvis in relation to the ribcage. We can do a movement assessment and we can assess the person when they walk. If they have trouble with anteriorly tilting the pelvis we can create some interventions where we anteriorly tilt the pelvis in relation to the ribcage as well as the femurs. The other thing to look at is the posterior tilting of the ribcage. This is interlinked to the anterior tilting of the pelvis; we want to see that the anterior tilting is happening together with the posterior tilt, as in figure 8.68. Effectively we want posterior bending in the whole spine, and a lift in front of the thoracic vertebrae, as in figure 8.55. If the body seems very stuck and it is hard to bring about a posterior bending in the spine, we can look at the feet and see if we have any movement of pronation available in the feet. When the rear foot anteriorly tilts it is easier for the pelvis to anteriorly tilt, as in figure 8.68.

8.72

Lastly, we have a situation in figure 8.73 where the pelvis is anteriorly tilted in relation to the ribcage and the femurs, as well as an anterior bend in the thoracic region, i.e. anterior tilt of the ribcage. This resembles what we described in figure 8.71, but this also includes more of the upper body. If the pelvis has a hard time at posterior tilting, we want to explore all the possibilities that we explored before (figure 8.57). This will give us the lift in front of the lumbar region, as in figure 8.58, but we also want to make sure that we get the lift in front of the thoracic region, as in figure 8.55. This lift is dependent on the pelvis journey from posterior to anterior, but the main tendency in this pattern is usually to get the pelvis stuck in an anterior position, therefore it is usually helpful to try to accomplish the posterior bending in the thoracic region without anterior tilting the pelvis.

8.73

Notes

1. I have been working on Gait Made Advanced since 2011. Gait Made Simple should be seen more as a pedagogical introduction to Gait Made Advanced.

2. If we want to make the concept of posterior tilt of the ribcage as it is talked about in the context of Structural Integration a bit more nuanced, we can say that it is the lower part of the ribcage that posteriorly tilts, and that the upper part actually anteriorly tilts. This also involves an anterior bend in the spine or flexion.

3. The whole spine seems to anteriorly shift (and extend) according to Thorstensson et al. (1984). Anterior shift seems to be 1–1.5 cm at C7 and 2–2.5 cm at L3.

4. The lift here is in relation to absolute space itself, or if you want, the ground/floor.

5. The movement of the sacrum is, from my point of view, clearly paramount when it comes to how the body functions in gait; however, the complexity (and uncertainty) of this topic makes it inappropriate to discuss it in this context. It will be further examined in Gait Made Advanced. But if we want to make things simple, we can just think about the sacrum as if it anteriorly tilts in relation to the ilium when the whole pelvis anteriorly tilts (this means that the ilium posteriorly tilts in relation to the sacrum), making it into a double spring system that can load up the system and shock absorb.

6. The navicular also medially tilts in relation to the talus.

7. Brockett and Chapman (2016).

8. Bozkurt et al. (2008).

9. Isman and Inman (1969).

10. This is, of course, a generalization, and every person has his uniqueness, where, for example, rotations in the bone structure can make a difference in the relationship in the transverse plane. If we make careful, detailed movement assessments we have a better chance to know what movement is related. Some people, of course, have more or less of this pattern in different areas, or contradictory patterns in some areas but not others, and sometimes there is a need to do a more nuanced description than this.

11. This "false pronation" differs substantially from "real" pronation; in the latter we have a proper load up of the soft tissue and a response from the body with resupination, while in "false" pronation the tissue just tries to hold on in an eccentric contraction with no recoil or resupination.

12. I consider gait assessment to be essential but usually not enough to get a full picture. In my practice, I use spring testing extensively, as described by Hesch (2015), together with other forms of assessments to get an exhaustive understanding of what might be going on.

References

Bozkurt M, Tonuk E, Elhan A, et al.; Axial rotation and mediolateral translation of the fibula during passive plantar flexion. *Foot and Ankle International* 2008, 29:502.

Brockett CL, Chapman GJ; Biomechanics of the ankle. *Orthopaedics and Trauma* 2016, 30(3):232–238.

Hesch J; The Hesch Method; integrating the body, recognizing and treating inter-linked whole-body patterns of joint and dense connective tissue, and reflex dysfunction. Workbook. Aurora, CA: Hesch Institute, 2015.

Isman RE, Inman VT; Anthropometric studies of the human foot and ankle. Biomechanics Laboratory, University of California San Francisco Medical Center, San Francisco. *Bulletin of Prosthetics Research* 1969, Spring:97–129.

Thorstensson A, Nilsson J, Carlson H, Zomlefer MR; Trunk movements in human locomotion. *Acta Physiologica Scandinavica* 1984, 121:9–22.

CHAPTER 9

Ensomatosy

Linus Johansson

For the many millennia, the human race has been on a journey of great cognitive and cultural development. On this journey we have tried to understand and explain the world around us and the more we have developed, the more advanced the explanatory models have become. This book is one more effort to make the picture clearer and more diverse.

Parallel to the development of our cognitive state and view of the world we have also striven to make the world a less hostile and demanding place to live in, an aspiration that has, over time, led us increasingly to abandon the chaotic, ever changing, living world and move into a man-made, constructed, "dead" world.

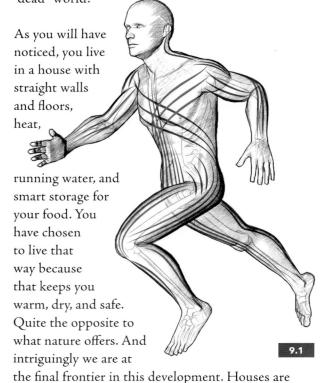

As you will have noticed, you live in a house with straight walls and floors, heat, running water, and smart storage for your food. You have chosen to live that way because that keeps you warm, dry, and safe. Quite the opposite to what nature offers. And intriguingly we are at the final frontier in this development. Houses are

9.1

smarter and less energy demanding, household appliances communicate with you, and everything is almost at the touch of your finger.

It all started when the first simple shelters and huts were constructed and raised by our forefathers, with the same wish as yours: to be warm, dry, and safe. When they created the first shelters they probably built them out of what they found around them, such as sticks and branches. The first ever homes and shelters may have been caves and natural cavities, but we could not choose the situation of these homes—they were where they were. Building our own homes in a preferred place could probably count as human beings' first wobbling steps away from being at one with nature.

Being able to build one's own home soon proved to be a successful strategy and ancestors who learned how to do this were the ones who survived. Through time and culture we have developed the art of creating homes, probably making it the most important, common, and unifying construction throughout the history of human existence. Thanks to sturdy, safe, and warm homes we could start to inhabit more of the surface of the planet and take our cognitive development even further.

New conditions create new opportunities. Well built shelters in combination with agricultural development suddenly unlocked something rare and valuable, namely time. We no longer needed to spend long periods of time looking for somewhere to shelter or looking for food to eat. We could stop using so much of our energy for these tasks and focus on new things. With a roof over our heads and plenty of food we could start to develop our cognition even more and parallel to that we could grow and expand our culture making it even more nuanced and diverse.

One big part of our culture has always been the effort to understand more about the world around us, and especially about ourselves, both from a philosophical perspective and most certainly also from the perspective of our human form and function. The question of what we are, how we function, what we are made of, and what resides inside us has always fascinated us. Looking back at how this was interpreted, we can see that our forefathers searched for answers using references from the world around them.

First there was the organic connection, which in many cultures is still one of the major explanatory models of the human form. The body was thought to be made up of different elements such as earth, fire, water and air. This is not surprising as these four elements were the most important and powerful components of life. They would also be the most obvious explanation of what the human form was made of.

However, through time, we have always taken our development further and reached new states of cognition. Parallel to this we have created a more complex way of life and more complex explanatory models to go along with it. We also made a shift in the interpretation of the human form in relation to its construction and function. Organic and nature-based ways of explaining the body became less important and we instead started to reference the body to the new and "dead" man-made world, the world that we had constructed out of "bits and pieces."

The idea that the human form was constructed out of small parts joined together was a much more fitting perspective compared with how we now lived, viewed, explained, and created the world around us.

Our evolution had taken us from the first small shelters to the solid and large buildings we now live and work in. Be it a simple shelter in the woods or a skyscraper in Manhattan, both are built up from parts connected and stacked on each other. We have created our world around us out of bit and pieces, why would our human form, or any other part of the organic world, be any different?

It is clear to see that the human mind has shaped the world we now live in, and at the same time that the world has shaped the human mind. When our predecessors for the first time set out to create a "map" of the human form, to create anatomy, their view was certainly colored by the notion that the world was built out of bits and pieces.

One should, however, be aware of one thing—you will always find what you are looking for. If you have the intention to go looking for bits and pieces, then bits and pieces you will find. Change your intention and you might find something else. However, the intention that the first explorers had was to find bits and pieces and so the first conceptions of human anatomy were created based on this intention.

We must be grateful for these perspectives and insights that our predecessors have given us, for without them we would never be where we are today and would never had come to the realizations that we now have. Hopefully we will also be mentioned in the future as "predecessors" and the "truth" that we present will also be revised and reformulated.

What we now know, and hold as true, is that we are not built up out of bits and pieces. The human form grows in the womb from a single seed to be an indivisible and completely connected unit, just like any other biological phenomenon. We also know that it was when we took this indivisible unit apart that we made a great cognitive error, at least when it comes to understanding human function.

As our cognition and culture keep on developing our intentions also develop and change. In fact they have always been in a process of changing—and will always keep on changing. What one sees when looking at a snapshot of human development is just one stage of transition into the next cognitive state. Nothing is ever still, everything is always moving.

To see and appreciate the human form as one indivisible unit is the main feature of the transition we are in now.

More and more body workers, personal trainers, and therapists are trying to see the greater, more holistic picture. The path toward this insight has been cleared by many before us. We acknowledge those before us who opened our eyes to the idea and whose discoveries we build on, and we are ever grateful to them. To be a successor is to stand on the shoulders of giants.

Our intention with this book, and specifically this chapter, is not to give a new and perfect idea and picture of the human form, but to be a part of the transition, to contribute to the ever-growing cognition. We are attempting to create yet another perspective of the "truth" and perhaps become predecessor to someone else, someone that in the future can make the picture even more clear.

Anatomy

The word anatomy comes from the two Greek words ἀνά ana "up" and τέμνω temnō "cut," forming the word anatemnō, meaning to cut up. Anatomy is what is called methodological reductionism: the attempt to explain the world around us in ever-smaller entities. The description of anatomy could therefore be the creation of something that does not exist by destroying something else that is already in full existence. Julian Baker gives his insights and great knowledge around this in Chapter 19.

One can argue that in a living breathing human being there is no anatomy. It is when a body is dissected and cut into pieces that we describe its anatomy. Anatomy can only exist in a dead body or theoretically in our minds, never in a living organism. This is why it makes it so difficult to use anatomy as an educational tool when seeing things from a more holistic approach.

This might be easy to say and even easier to dismiss; however, anyone who has ever taken part in a dissection knows what you encounter, and it is not bits and pieces. Taking a human body apart is a very demanding task and is not at all like pulling apart Lego bricks from each other. What first meets you when you approach a deceased human form is the unified covering of the skin, designed to keep the inside of the body safe and secluded and at the same time be one of the greatest and most sensitive organs of the human form.

Even in a deceased body the resistance that the skin gives to protect what lies beneath is remarkable. Within minutes of starting to cut into the skin, the razor sharp scalpel becomes blunt, making the work of removing the skin from the underlying tissue very hard work.

Once the skin is removed a beautiful world of colors and patterns is presented. The sense of layer upon layer of unified indivisible tissue lie beneath the skin, reaching in all directions and seamlessly passing throughout the entire form of the body. Seeing this, one cannot deny that we are, from the surface of the skin all the way down to the bone and from the head to the toes, connected as one indivisible unit.

The deceptive side of this is that the human eye is very sensitive to shapes, forms, lines, and colors. We are always looking for differences and contrasts, which make us very prone to separate, segregate, and reduce.

It is not hard to see this in other contexts. You just have to look at yet another system to see this very clearly, namely the society around you. You will probably see more segregation than integration going on between different groups of people trying to live and function together in our society. This is a harsh and unpleasant reflection of how the human mind works. We are, unfortunately, notorious methodical reductionists in all layers of life and tend to look at all systems in this way.

This is a clear and obvious weakness in human cognitive capacity—we need to separate to understand. We need to focus on one thing at a time to be able to see, explain, and communicate it to others. We always have difficulty trying to understand the bigger picture if we cannot view it in smaller pieces.

This holds as true for understanding the human form as it does for other kinds of understanding. We constantly search for borders, clear structural boundaries, and differences in color and shape. We need these contrasts to let our mental scalpel pass over and through the tissue to segregate, reduce, and diminish it, so we are able to understand it. The question is if we can ever truly understand the human form if we keep treating it this way?

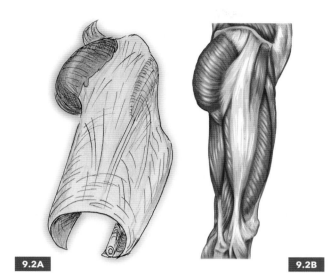

9.2A   9.2B

One good practical example of this is the iliotibial tract or IT band. This is described in the established literature as a thick band that passes over the outside of the leg, from the hip down to the knee (figure 9.2B). It is presented as a clear and separate piece in most anatomical atlases and is one of the many structures that bear an unjustified burden in the therapeutic world and especially in the fitness industry. It is a structure that is often blamed and tortured, rarely loved or appreciated.

> *"We are not just dissecting bodies here, we are dissecting our knowledge, we are dissecting our language and we are dissecting our beliefs."*
> —Julian Baker

These words, spoken by Julian Baker during the first dissection day in Nottingham on the third of April 2017, truly became a keystone in what we learned and understood during that week.

When looking at the skinned thigh in the dissection lab it is strikingly clear that the IT band is not a separate piece of tissue running from the hip to the knee, as shown in the books. The IT band is a thickening of the layer of fascial tissue that encircles the leg (figure 9.2A). The thickening is an adaptation in the fascial structure due to the high demand of transmitting and handling forces on the outside of the body when moving in the gait cycle through gravity.

The IT band therefore cannot be removed with one's bare hands. The only way to be able to state that there is an IT band is if we decide where it resides and with our scalpel cut it out from the structures surrounding the thigh, thereby taking it out of its context and destroying all it is connected and related to. What happens is that we create the IT band and at the same time completely destroy the complexity.

On page 89, fascia lata and the iliotibial band, Gary Carter elaborates and discusses this in greater detail.

Once again we see that we have to destroy and segregate to try and understand, as our limited minds are not capable of taking in too many complex parameters at the same time. One can argue that separating and segregating is just for educational purposes and that the intention is to show the parts of the whole. The problem is that by cutting and creating the IT band, or any other structure, and giving it an origin, attachment, and function, one also limits the understanding of its purpose. The scalpel does not just cut the flesh of the body, it also cuts the mind, limiting even more how we see and understand the bigger picture. As all structures in the human form have connections in all directions and depths, we limit and diminish ourselves by creating them. A classic swing and miss.

Looking at the outside of the leg again, the thickening of the structures is an integrated part of the complex and vast three-dimensional structure that unifies the body. All structures receive and transmit forces of activity and compile them to create perfect and energy-efficient movement. Therefore the IT band cannot be just a lateral band that does one thing and one thing only on the outside of the leg. It is just one of many continuous alignments of tissues in the vast complexity of the human form.

> *"It comes from nowhere, and nowhere is everywhere."*
> —Martin Lundgren

During one of our many lectures together Martin presented these winged words. One of our students had asked the classic "where does it come from" question. In all its simplicity, the answer is in itself very deep and complex.

One needs to be aware that seeing the human structures this way can be rather demanding, for the number of ways all our structures can interact is infinite and cannot be reduced to one direction or one plane of motion. What we need to realize—and this goes for all human properties—is that our limited minds may truly never be able to comprehend the vastness of its complexity; the only thing we can do is just to acknowledge it. It is a great paradox that we just have to live with.

On the other hand, if instead of trying to understand the complexity we just choose to acknowledge it we can reach other possibilities. In this book we try to give you the principles to be able to come to such an acknowledgement. We claim that if you have solid and well-grounded principle you can let those principles be the connection into the complexity. This way you do not have to understand and explain everything that happens; you can just appreciate the existence of the complexity and the fact that it will, through the principles, aid you when working to create development in human function.

In the light of all this, the question becomes clear; can we create an explanatory model without reductionism? Can we create something that can give us another view of the human form and function and something that provides the means to let us just acknowledge the complexity without the need to explain it?

The answer would be "Yes" if we use a perspective that utilizes the principles to uphold the explanatory model. However, can we really create an explanatory model that could cover all the different perspectives and visions of how the human form moves and functions? The answer to that would be "Probably not!" The next question would be, do we need to? The answer to that is, "Definitely not!"

Why? Because the same goes for classic anatomy; it is a good explanatory model that gives one particular view of the truth and works perfectly well from some, though far from all, perspectives. The same goes for our explanatory model; it will fit some perspectives but not all.

The new expression

From the perspective we present in this book classic anatomy is insufficient in its ability to fully accompany our explanatory models. We do acknowledge classic anatomy and understand that it is a part of our heritage and culture and that it fits many other perspectives. It is also a big part of the language we use for structures and when speaking of the landscape of the body. However, in our perspective the names of different "parts" are viewed as being more like postcodes and we use them to communicate and to orient us around the human form.

We would like to underline that we do not diminish classical anatomy; however, we must state that we cannot use it fully as intended in our perspective. As mentioned earlier, the very word anatomy comes from the act of cutting something up into smaller pieces, with the intention of them being used as an explanatory model for human form and function. In other words, anatomy is taking apart what we are, but do not understand, in order to rebuild it into something we understand, but not something we are.

But what if we did it the other way around? Rather than bits and pieces, what if we tried to understand the body as a whole, through function, and then addressed topics and more local phenomena in the structure without losing the idea of it all being connected?

The word anatomy is unfortunately laden with preconceptions of the human form because of its reductionist heritage. To be able to move forward, and perhaps beyond, we therefore need a new expression to work with, an expression that can contain the new perspective and the new insights.

As the word anatomy is derived from the beautiful Greek language it is right to honor our heritage and the women and men that preceded us and also find this new expression in the Greek language.

The criteria for the expression were as follows:

+ It should not to be a made up word.
+ It should have an actual meaning in Greek that would be in line with the concept.

+ It should be fairly easy to say and spell.
+ And it should have a phonetic resemblance to "anatomy."

One day, in May of 2018, at the Mad Hatter Cafe in Brighton, the new word was discovered; it is *ensomatosy.*

> * Ensomatosy from the Greek word
> Ενσωμάτωση – Ensomátosi – meaning
> integration, embodiment or/and
> incorporation.
>
> **Definition:**
> ** Ensomatosy is the interpretation and
> study of the internal relationships between
> structures in the human form and the
> external relationship the human form has, as
> a whole, to gravity and movement.
> ** Ensomatosy is a parallel perspective to
> anatomy, describing the human form as an
> indivisible and fully interconnected unit.
> ** Ensomatosy is an interpretation that states
> that understanding and explaining human
> movement and function only can be done
> when the body is viewed and addressed
> as a system.

Ensomatosy is not supposed to be a substitute or replacement for the word anatomy. Anatomy is a very clear and established perspective that has a given place in this field of work. Ensomatosy is formed to be a parallel perspective to anatomy. Ensomatosy stands at the other end of the explanation scale to show that there can also be a non-reductionist view of the human form, i.e. a holistic view.

We acknowledge that the need for a word and expression of this sort has been around much longer than this book and that other expressions of this sort might already exist and be in use. However, we have not found any expressions that matched our needs and therefore we stated our own.

We have also replaced the "i" in the original "ensomatosi" to a "y". The purpose is to distinguish it from its original form that has an actual meaning in Greek and is used in many other contexts. With the "y" at the end ensomatosy instead becomes a unique expression, still with very close connections to its heritage and original meaning.

We hereby declare the expression ensomatosy, as a defined expression, free to use by anyone that finds it useful in their line of work.

Fascia lata and the iliotibial band

Gary Carter, from the Human Fascial Net Project, Germany

> *"It is not possible to distinguish the medial and lateral border [of the IT band] because it continues into the fascia lata, appearing just as its longitudinal reinforcement."*
> —Carla Stecco

The topography and descriptions of the iliotibial band may be of one of the most misunderstood structures of fascial anatomy due to the many illustrations that show the iliotibial band as a separate structure.

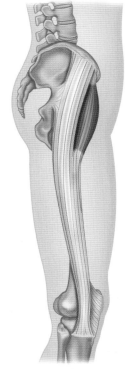

The iliotibial band (ITB) does not actually exist as an independent structure in the living form, rather it is a lateral reinforcement of the fascia lata which envelops the entire thigh. This reinforcement comes as a result of loading and does not exist in babies before they learn to walk.

Dissection studies of the iliotibial band in bears, gorillas, orangutans, chimpanzees, and various quadrupeds, concluded that although all quadrupeds have both gluteus maximus and tensor fasciae latae muscles, they do not have an iliotibial band. The consideration here became

9.3

that the presence of the iliotibial band becomes a stabilizer for the lateral aspect of the knee, which becomes essential for the erect posture.

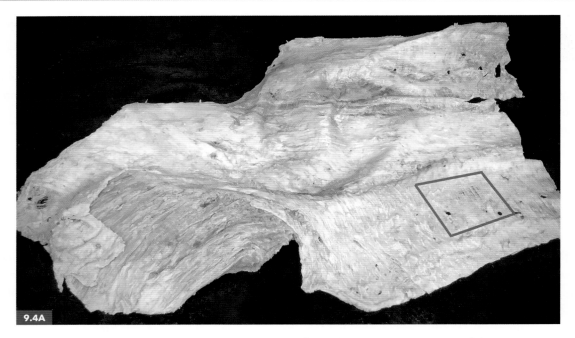

Dissection of the fascia lata with close up marked to show the fiber directions characteristic of this tissue.

Image reproduced with kind permission from the Fascial Net Plastination Project (https://fasciaresearchsociety.org/plastination).

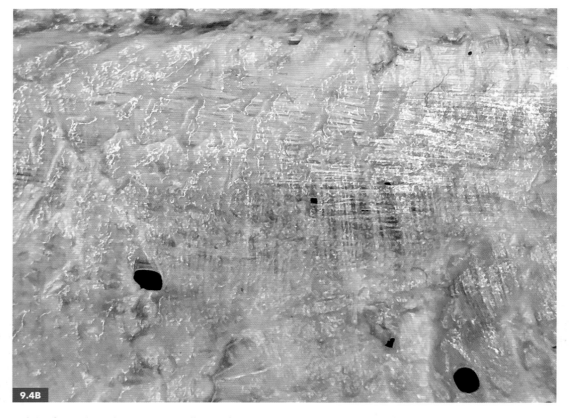

Close up of the fascia lata showing the collagen fibers typical of this fascia which are a reinforcement of the tissue caused by repetitive strain due to force transmitting through the fascia.

Image reproduced with kind permission from the Fascial Net Plastination Project (https://fasciaresearchsociety.org/plastination).

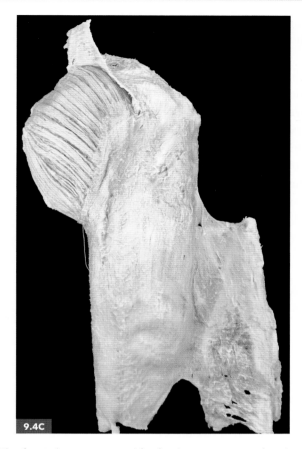

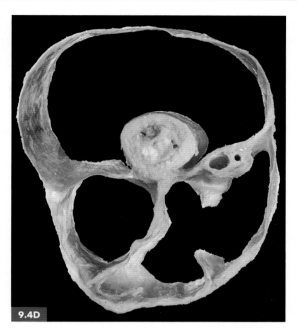

The finished plastinated 5 cm cross section of the thigh. The fascia lata forms the outer ring with the femur in the center and the fascial septa dividing the gross compartments.

Image reproduced with kind permission from the Fascial Net Plastination Project (https://fasciaresearchsociety.org/plastination).

The fascia lata positioned for final preparations after the plastination process.

Image reproduced with kind permission from the Fascial Net Plastination Project (https://fasciaresearchsociety.org/plastination).

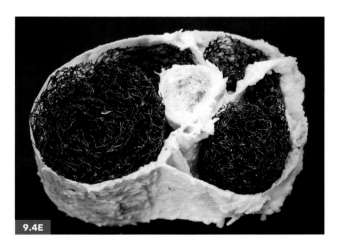

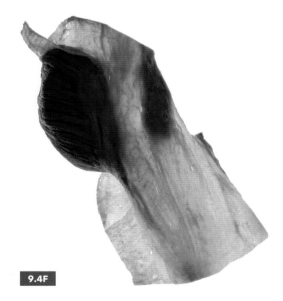

A 5cm cross section of the thigh as a 3-D fascial structure. The entire fascia lata forms the perimeter, with the femur in the center, and the larger septa dividing the gross compartments. The dissection is positioned with a mesh netting to hold its shape and form before plastination.

Image reproduced with kind permission from the Fascial Net Plastination Project (https://fasciaresearchsociety.org/plastination).

The final, plastinated fascia lata. The gluteus maximus is exposed with a fascial section folded back on the upper left, showing the underlying muscle. The tensor fascia lata is visible within the tissue in the upper right. The continuity of the fascia lata and the ITB can be seen in the bottom half of the image with the ITB appearing slightly darker because of its thickness from an increase of vertically-arranged collagen fibres.

Image reproduced with kind permission from the Fascial Net Plastination Project (https://fasciaresearchsociety.org/plastination).

The ITB is likely to have formed from standing upright, created from longitudinal loading and motion, bringing tensional support and elastic storage potential along with efficient containment of the ongoing muscular hydraulic forces. This has then created the vertical appearance to the tissues showing possible lines of force along with lateral and multi-directional patterns in the tissue giving it its distinctive appearance.

The fascia lata is a surrounding fascial sheet of the deep fascia around the thigh to the lower leg; this includes the quadriceps group, hamstrings, adductor group, including the sartorius and gracilis, which are all interwoven into this fabric.

At its uppermost region it includes the tensor fasciae latae, gluteus maximus, and connected deep to its surface is the gluteus medius. At its lower region the fascia lata continues to the crural fascia of the lower leg; these tissues also fold into and continuously envelop the myofascial units of the lower leg.

The densities and textures of these tissues vary in different regions of this sheath. The fascia lata laterally at the thigh is thicker and becomes thinner and more elastic medially and posteriorly; toward the lower leg it blends with the periosteum of the tibia also enabling attachments to the tibialis anterior from its deep surface. It appears thinner posteriorly on the lower leg to allow for expansion of the calf group, gastrocnemius and soleus also appearing thinner as it surrounds the gluteus maximus.

The iliotibial band also appears thicker not only from the upright loading but also from dividing and delving deep between the vastus medialis and biceps femoris reaching the posterior surface of the periosteum of the femur at the region called the linea aspera, meaning "rough edge"; from here it continues and divides through the adductor group, further dividing to create the intermuscular septa of the entire adductor group, where it regroups with the fascia lata medially both surrounding and enveloping the thigh.

Facial plastination of the fascia lata as a sheath and cross-section

This structure has finally been imaged and created as a 3-D structure using dissection and plastination methods by the Fascial Net Plastination Project at Gunther von Hagens Plastinarium in Guben, Germany, and was displayed for the first time at the World Fascia Research Congress in Berlin 2018, as shown in figure 9.4A, B, C and F.

The process of creating this piece was achieved by removing the skin and superficial fascial tissues, which include the adipose tissue. The entire muscular sleeve

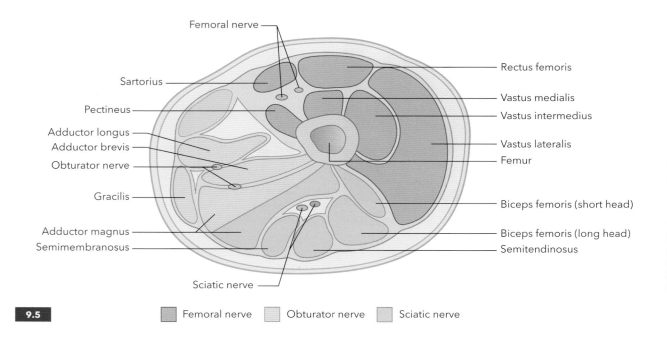

9.5

Femoral nerve · Obturator nerve · Sciatic nerve

was carefully removed from the femur with a midline cut along the level of the sartorius and from just above the knee, the entire gluteal region was also included. With the deep surface facing upward, the procedure was to remove all muscular tissues from the fascia lata, whilst saving some septa. What remained was the entire fascia lata from above the knee to the pelvis, the deep septa of the medial thigh, including the deep femoral artery and the septa between the vastus lateralis and the bicep femoris. The only muscular structures that remained intact were the gluteus maximus and the tensor fasciae latae. This then went through the various procedures of treating, positioning, preparing, and plastination to create its final result.

The plastinate you see illustrated here is the fascia lata, the deep fascia of the right thigh, from just above the knee inferiorly to the iliac crest superiorly, with the gluteus maximus and tensor fasciae latae muscles still attached and intact. The gluteus medius, vastus medialis, and vastus lateralis were all removed from their attachments to this covering. The fascia lata is continuous with the fasciae of gluteus maximus and medius and the tensor fasciae latae muscles. The incision for this cylindrical structure was made by following the path of the sartorius muscle. Running vertically on the interior of this plastinate, you can see what remains of two fascial septa, which were previously attached to the femur forming compartments of the thigh. One of them still holds a section of the deep femoral artery enveloped in the septa.

The other view is a 5 cm cross-section of the mid-thigh

To provide a good view of the deep fascia of the thigh, the fascia lata, with its septa, this plastinate of a 5 cm

cross-section of the mid-thigh was created, as shown in figure 9.4D and E. The superficial fascia and the muscular tissue have been completely removed to reveal the discrete compartments that are formed by the deep aponeurotic fascia: the anterior, medial, and posterior compartment with the femur at the center. It is easy to see how the septa provide structural stability to the system. The septa are connected, like spokes of a bicycle wheel, to the outside "tire" of the external fascia lata and firmly anchored to the central "hub" of the femur.

You can also see on the left lateral side of the image where the iliotibial band would appear, to "create" an iliotibial band requires making a decision to cut the fascia lata at a specific region to make it a separate structure.

References

Kaplan EB; The iliotibial tract; clinical and morphological significance. *Journal of Bone and Joint Surgery* American volume 1958, 40-A:817–832.

Stecco C; *Functional Atlas of the Human Fascial System*. Edinburgh: Churchill Livingstone, 2015.

Stecco A, Gilliar W, Hill R, Fullerton B, Stecco C; The anatomical and functional relation between gluteus maximus and fascia lata. *Journal of Bodywork and Movement Therapies* 2013, 17(4):512–517. http://dx.doi.org/10.1016/j.jbmt.2013.04.004.

Wilke J, Engeroff T, Nürnberger F, Vogt L, Banzer W; Anatomical study of the morphological continuity between iliotibial tract and the fibularis longus fascia. *Surgical and Radiologic Anatomy* 2016, 38(3):349–352.

CHAPTER 10

Ensomatosy visualized

Linus Johansson

In this chapter we present an interpretation of a visual expression of ensomatosy. Please note that this visualization is much more an artistic expression than a scientific absolute.

This visualization of the human form was initially created to support the learning process during an apprenticeship program, originating from the perspective presented in this book. The need to connect the newly acquired knowledge and insights with something visual proved crucial in aiding the process for the students. The interpretation from classic anatomy failed to give the support needed from this perspective, and the development of a new visual expression was needed.

The main purpose of this visualization is to give the impression that everything in the human form is connected and indivisible and to show, once and for all, that there are no bits and pieces. Therefore, the lines of the illustration pass over the all common borders and boundaries and unite what has for so long been "taken apart." The color gradient equally helps to diminish the borders and to give the viewer the impression of a variation of form, without segregation.

The lines do have a resemblance to the old muscle maps and this is not a coincidence. It is obvious that we do have "lineage" in the body and that it is truly connected to our function. However, the lines in this visualization are not intended to be the full representation of the muscular fibers in the body, nor of the fascial structures.

The lines are instead intended to express three main aspects:

1. First, when a body moves through gravity these paths will be eccentrically loaded in different patterns and combinations, making them the paths of load.
2. Secondly, the lines are, in this perspective, "the most common paths" that our hands travel over and across when working to develop a body's movement potential, making them also the paths of development.
3. Thirdly, the paths are intended to be a visual aid to clearly see that everything is connected, making them also the paths of integration.

Together, these paths of load, development, and integration create a new visual expression of the human form. This, in combination with the principles of this perspective, creates a new way to see and explore human movement.

Let us dive into each aspect a little deeper.

Paths of load, development, and integration

The human body is "designed" to be able to move in as energy-efficient a way as is possible. One of the properties needed to be able to do that is loading the tissues eccentrically to create what is called elastic recoil.

We will just touch lightly on the terms "eccentric" and "elastic recoil" in this explanatory model. What we would like to underline is that these two terms are also interpretations of the vast complexity of human function, meaning that we can recognize them and name them but never truly explain them and foremost, never segregate them from the complex movement system. All we need to do, and can do, is just acknowledge that we can create this interpretation and use it in our practice.

When we move through gravity, especially during gait, it is the intricate structure of the human form that holds us afloat. In this floating structure the bones move in relation to each other and elongate the tissues that extend between them. This elongation loads the tissues with movement energy which in turn, when the movement alters and the relationship between the structures changes, creates the elastic recoil. Many of the bones in the body have the same kind of property and can be loaded and also work as springs in combination with this elastic recoil.

The first aspect with the paths of visualization is to show the representation of the most common way that these tissues get elongated and loaded when the body moves through gait. Although this visualization is two-dimensional we recognize that all tissues always load in all three planes of motion at the same time when we move.

The view of the paths of **load** seamlessly then interconnects directly to the second aspect, the paths of **development**.

In Chapter 5, we speak of how each individual has a theoretical optimal body that we can aim and strive for when developing their abilities. The way to find a person's potential and abilities is to do a structural assessment while they are standing still in gravity. What we look for is altered relationships between structures, as also discussed earlier in Chapter 5. We then put these alterations to the test to see if the movement or function is lost or still present.

Now, say we have found an altered relationship between two structures, e.g. the pelvis and chest, as shown in figure 10.1. What we see when we look at the person is that the pelvis has tilted to the left in relation to the chest. The distance between the left side of the pelvis and the left side of the chest is greater than on the right side, where the structures are closer. When we then test to see if there can be a separation between the pelvis and chest on the right side we find that this is difficult for the person.

To relate this into our visualization, if we see two bony structures that are in a closer relationship to each other we understand that the tissue stretching between these structures has altered in length, and we describe it as held short. The same goes for the other way around. If two structures are taken too far away from each other we can describe the tissue stretching between them as held long. In our example we would describe the tissue on the right side of the torso to be held short and the tissue on the left side to be held long. Please note that the expression "held short" is not the same as the more common expression "tight muscle" or "short muscle." These are two completely different aspects based on two completely different interpretations.

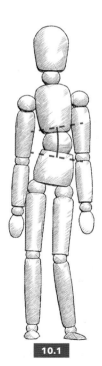

This is where the ensomatosy visualization comes into use. We can now use it as a map or chart to determine which paths are affected by this altered relationship and in turn see how these paths connect further into the movement system.

When you look at the dummy in question you can see that more things have happened than just a tilt of the pelvis to the left (figure 10.1). This is always the case. There is never just one alteration of relationships in a human form. When something gets compromised, for any reason, there will always be a domino effect of altered relationships up and down the movement chain.

10.1

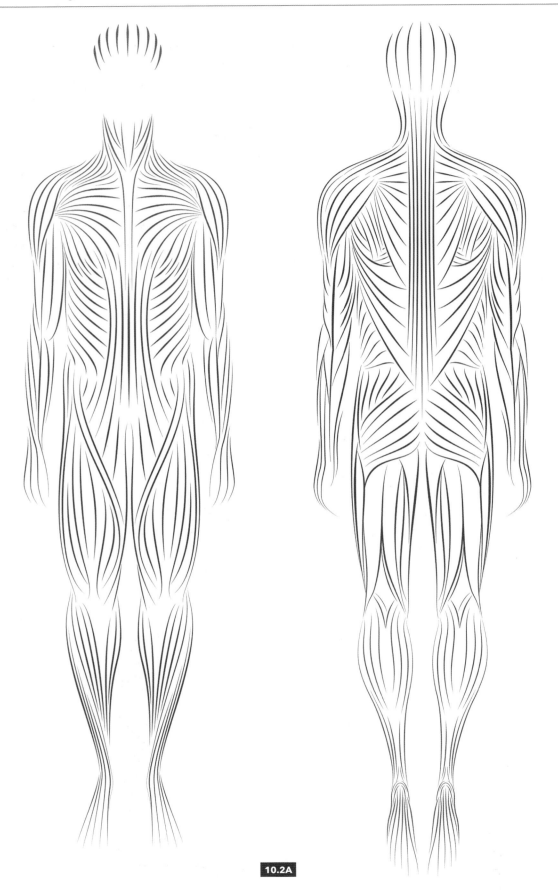

10.2A

10.2B

Ensomatosy visualized

The ensomatosy illustrations are made to show the most common paths that eccentrically loaded the tissues in the body in different patterns and combinations, making them the paths of load. The illustrations also show "the most common paths" that our hands travel over and across when working to develop a body's movement potential, making them also the paths of development. Finally the ensomatosy illustrations are intended to be a visual aid to clearly see that everything is connected, making them also the paths of integration.

To express this altered pattern in the ensomatosy visualization model the different affected areas are highlighted. In this case we color in the tissues that are held short with purple and the tissues that are held long with green.

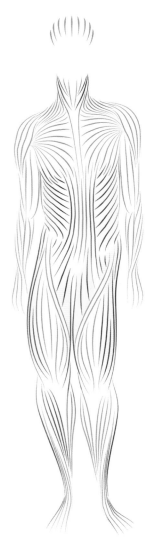

This illustration shows how a structurally held pattern can be visualized using the ensomatosy model. The purple represents where the tissue is being held short and the green represents where it is held long.

In this example we use the visualization to describe the pattern in our imaginary client, but we could also use this principle to describe any circumstance, pattern, or movement in a human form. Say we took a picture of a person in a sporting event, as in figure 10.4; we could mark out which tissues were held short and held long, in function, for that fraction of time. However, this would not be because of a

10.4

This illustration shows how the functional relationships in a snapshot can be visualized using the ensomatosy model. The purple represents where the tissue is held concentrically short and is in the recoil phase of the movement. The green is representing where the tissue is being loaded and taken eccentrically long.

compensation due to structural altered relationships, this would instead be the illustration of the loading and exploding phases of the tissues when in full movement. We can illustrate this in a snapshot; however, the situation would be quite different just a second later, when the athlete continued the movement into a different pattern.

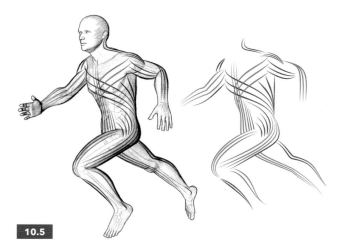

10.5

It is beautiful to still be able to see the shape of a person when we just illustrate the concentric and eccentric paths of their movement.

For our client, we see the held pattern that they are having trouble moving away from. Our mission is therefore to see them, just as we see the snapshot of the athlete, and try to understand how we can help them progress further into the next level of movement.

Therefore, we invite you to understand that what we see—and what the ensomatosy visualization represents—is always movement. This way of illustrating is yet another perspective to see and understand the complexity of the moving human form. It is free to embrace and utilize for anyone seeing the benefits of it.

Back to our imaginary client. To be able to do an **intervention** and to create development via this perspective we need to have a systemic approach and a holistic engagement. We need to invoke all the combined aspects of the person's form, from head to toe, from skin to bone, and from flesh to soul.

Therefore it is important to understand that when we meet a client, instead of trying to determine what is stiff, weak, or insufficient, we instead let the ensomatosy visualization represent the connected potential in that person's body. This guides us to how and where we can work with a person to create development and functional change.

The color purple in this visualization thereby becomes the representation of potential! It is a visualized construction based on the combination of our findings in the visual assessment and the results from our movement tests.

The second color, green, also represents the lack of movement but for the completely opposite reason, as it is due to the fact that the structures have been taken so far apart that there is no more movement to execute, making the affected tissues held long.

The reason why these tissues become held long is due to the fact that if one movement happens in one place, in a structural relationship, there will always be the opposite movement somewhere else. This is one of the basic properties of human locomotion. Hence held long and held short always exist in a synergy.

Intriguingly, these two aspects appear to be quite different when one studies them objectively, and they are. On the other hand, when experienced subjectively they feel in most cases the same. A person trying to create a movement over a tissue that we have assessed to be held short, would perhaps argue they feel "stiff" because the wished for movement is demanding to execute. The person would also express more or less the same feeling when trying to elongate a tissue that is held long as it has already reached its maximum length and can't be taken much further. The feeling of not being able to move freely is often underlined in both cases.

This experience is due to the fact that most people do not have a body awareness that lets them access these insights to the different relationships in the body. Most people walk this earth trying to experience their being as little as possible. This creates a numbness and means structural differences are all experienced more or less in the same way.

A confusing and somewhat depressing reflection is that even highly trained persons have surprisingly low insight into this and little ability to distinguish these differences. This may be partly because they study methods and not principles.

Our mission, when we come across this state of affairs, is to take the opportunity to create development and increase awareness within the person. We do this by restoring the dynamics in the relationships in the person's movement system and inviting them to meet themselves in motion. Creating actual physiological change and development in the flesh and tying it together with the psychological expansion of the mind. Trying to make the person come as close to the evolutionary utopia, "… to move freely in body and mind …"

How to accomplish this then comes down to the choice of intervention. Manual therapy in combination with movement and education is in general our preferred choice. Through our interventions we can utilize the body's complexity, intelligence, and desire to always become more energy efficient and reintegrate the dynamics between the internal structures. Subsequently we can then reintegrate the relations between the complete system, i.e. the person, and the gravitational forces of the earth.

However, how we methodically accomplish this is not as important or interesting as the fact that, via our principles, we were able to give a systemic explanation as to how the relationships were manifested in the person's form, i.e. understanding where the person is, the nature of the journey that lies ahead, and what is needed to be able to strive for the desired goal.

What the paths represent

As discussed, we describe the paths in the ensomatosy visualization model as the paths of load, development, and integration. However, these are theoretical descriptions and they become quite vague in relation to the actual illustration where the lines are clearly drawn over the shape of a human form.

While in theory they are the paths of load, development, and integration, in practice they are the representation of the human form, i.e. the skin, fat, superficial layers of tissue, deep investing layers of tissue, muscle fibers, the septum of the different muscular compartments, bones, ligaments, nerves, nerve endings, nociceptors, thermoreceptors, mechanoreceptors, veins, arteries, lymphatic system and viscera, and perhaps even the soul.

Therefore, when we engage into a human form with our intention, using our hands or movement, we must always acknowledge that we engage into an indivisible unit, where everything is connected; from skin to bone, from head to toe, and from flesh to soul. The path of movement integration therefore states that if we touch somewhere we touch everywhere and if we move something we move everything.

Our intention with sharing the ensomatosy visualization model in this book is for you to be able to connect it with the principles given in this perspective and to perhaps inspire you to a more holistic approach when working with your clients. The ensomatosy model is also useful to communicate the holistic principles with our clients and let them connect with the holistic principles via a visual expression.

Nevertheless, however unifying the ensomatosy visualization model is, it still contains one great bias. It is lacking one very crucial aspect that, unfortunately, can never be included in any visualization. It is also perhaps the most important and at the same time most challenging part of a successful movement integration. It is of course the human mind.

We know that a lot of human disorders, including pain, can be derived from the human mind and that movement and manual therapy do not solve all problems. Some issues can only be solved using psychological treatment. However, as in all human aspects, there are no clear lines, only gray areas. Therefore, as movement and manual therapists, we can never exclude the human psyche from the equation if we want to be sure to have a successful treatment or to reach the developmental stages we aim for.

Neither the concept of ensomatosy nor the visualization can include the human mind in its definition and structure. However, when ensomatosy is fused together with our other principles and formed into the concepts of systemic movement integration the human mind is very much included.

This topic will be further addressed and integrated in Chapter 12.

CHAPTER 11

To study ensomatosy

Linus Johansson

"To study ensomatosy is to dedicate oneself to a never-ending journey of exploration and to learn to trust the intelligent body."

Few of those who study classic anatomy are given the chance to take part in a dissection and truly look inside the human form. Most students learn anatomy through a book, studying the names of the muscles, their origins and insertions, and of course the given function the muscles have. All this is taken from studies of dead and embalmed bodies where someone's ideas, with the help of the scalpel, have created the established "truth."

We have already discussed our take on classic anatomy and how we believe that it is flawed, especially when seen from the perspective of ensomatosy. We do not claim anatomy to be wrong; we just think that it is strange. Strange because we all have full access to the living human form—it is all around us and it is free and easy to access unlike an embalmed body in a morgue. We ask how come we don't study the living form, as this is what we will be working with in our profession. It seems strange to study the dead when we are supposed to work with the living. Julian Baker underlines the danger of comparing the living with the dead.

"It's a dangerous thing to make assumptions about function and movement, based on dissecting people who are no longer functional or moving."
—Julian Baker

One might claim that we do study the living human form in our established educational courses, but do we really? How many courses in therapy and movement do not first educate students in classic anatomy before going on to studies of assessment and function?

Very few students get the chance to understand human living function without first having to learn classic anatomy. Their minds are then filled with the reductionist idea that the body is made out of bits and pieces and when starting to treat and move the body they do not have anything else to relate to.

We encounter this a lot in our open educational sessions where people with earlier classic education participate. Students with a preconception of how the body looks beneath the skin have a much narrower space to grow in because classic anatomy has given them well-defined boundaries and borders to stay within. They also become more easily in conflict with themselves when trying to use a reductionist approach to learn the holistic idea. It takes a lot of extra energy and effort for them to see beyond and into more complex relationships.

So, what if anatomy and the interpretation it gives is not the complete truth? What if there are other ways to see and understand the human form? We know that we might be upsetting quite a few readers with our criticisms of established orthodoxy but we do hope you reflect on our reasoning and follow us to our conclusion. All paradigms are open to questioning—if they were not we wouldn't have any progression or development. The same goes for our standpoint, and we encourage you to question us and interrogate what we claim; that is a healthy and reasonable attitude.

Our reason for questioning the established paradigm is because what it all comes down to, in the end, is not the years of study we have done or all the degrees we have taken, it is the outcome for our clients that counts, and is the only thing that counts. The clients do not get any better just because you are highly trained; they get better because you see and treat them for what they are, a free and flowing entity and not a constrained and dissected idea.

With this book and the concept of ensomatosy we want to contribute to this interpretation of the "truth" and also challenge ideas and raise a number of questions. Can we learn to see and appreciate the human form for what it is, i.e. something that is living and not dead? Can we let go of reductionist concepts and still succeed? Can ensomatosy be the concept that we can use to embrace all these ideas? Can this give us other ways to reach more people and make them move and function better?

To answer these questions we challenge ourselves to prove them. In the apprenticeship program mentioned earlier, none of the students had done any anatomical studies before, nor had they undertaken any physiological or therapeutic studies. However, what they did have was a love for movement and an interest in understanding more about human form and function. Instead of undertaking classic anatomy studies from a book they were introduced to the studies of ensomatosy.

The study of ensomatosy is as much theory as it is practice. The practice is, however, much more easily accessible due to the fact that you need living and breathing bodies, unlike the practical studies of anatomy. However, there is nothing that says that you cannot study ensomatosy in a dead and embalmed body. It is not the act that counts; it is what you are looking for.

We have a lot of talented colleagues and friends who explore the human form with the scalpel and do not look for the classic bits and pieces. Instead they look for the connections and continuities. Two of them are contributing in this book, Gary Carter and Julian Baker, and one cannot mention this concept without also acknowledging the great work of Gil Hedley, Tom Myers and Todd Garcia. All are pioneers in this line of work and interpretation.

To be able to study ensomatosy there are three basic needs. First there is the need for a holistic principle. Secondly there is a need for an entry point into the

principle. Thirdly there is the need for a positive and creative mind. If we have all these, we can undertake the studies of ensomatosy.

How we study ensomatosy

The principle we present in this book is that the meaning of life is to "move freely in body and mind" and that we have evolved to walk on our two feet through a very specific biomechanical rhythm.

11.1

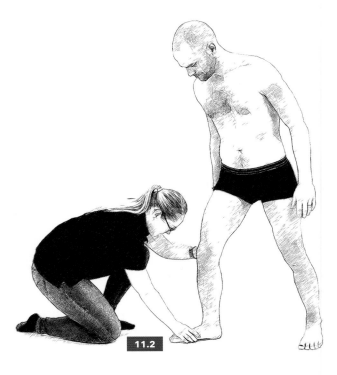

11.2

The entry point to this principle is the "two patterns," something we will elaborate on soon.

The positive and creative mind is the driving force in any walk of life, as it is in this. We speak of the power of the mind in Chapter 12. Using the principle to guide us and the two patterns to enter the principle in a practical way we explore by asking, looking, moving, feeling, touching, sensing, assessing, testing, and evaluating everything with the intention to find opportunity, potential, and development.

To learn all this is a great process in itself. We must be careful as we do so that we do not create a monoculture as studies of classical anatomy do. In anatomy the biceps will always attach in the same place and do the exact same movement, always. What we aim to do in ensomatosy is instead to create a polyculture where we acknowledge that we have not yet completely discovered and fully understood the human form and function. This opens up our minds to other possibilities and keeps the exploration alive.

We therefore, once again, underline that ensomatosy is more of an artistic approach than a scientific one. If anatomy, with its scientific and reductionist approach, can be placed at one end of a spectrum, ensomatosy, with its holistic and artistic approach, can be placed at the other end.

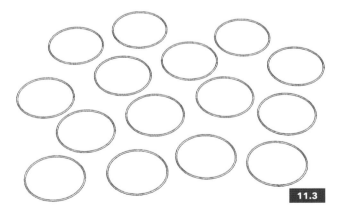

This illustration is a visualization of the reductionist and segregated approach to the human form. Each named topic is held within each circle and has no connection with the circle lying next to it.

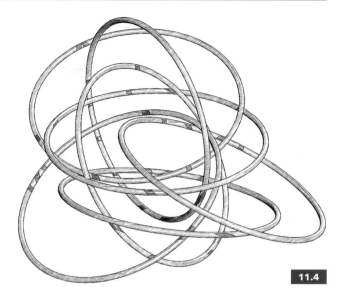

This illustration is the visualization of our holistic and integrated "truth." Everything is closely connected and deeply dependent on each other. Affect one of the circles and you will affect them all.

The two patterns

When trying to enter the vast and maze-like complexity of the human form we need a clear and wide entry point to come in from before we can narrow it down. We do this by creating two theoretical and rather utopian patterns, named and derived from the principles of the foot's ability to pronate and resupinate, as described in Chapter 8.

We name the patterns "the pronated pattern" and "the supinated pattern." The theory is that when a movement of pronation or (re)supination occurs in the foot, a particular chain reaction of movements will travel up the body and create a specific biomechanical rhythm based on these two movement patterns. This rhythm travels back and forth in the movement chain when we move on our two feet in the gait.

Each movement pattern can be described to start anywhere in the body and therefore equally create the rhythm downward into the foot. This is what constitutes this very specific loop of movements, up and down the body when we walk and run. It is this rhythm

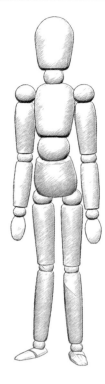

The pronated pattern's major structural features are the pronated feet, the relative inward rotation in the bones of the legs, and the relative tilt between the pelvis and chest creating a long tummy and a short back.

The supinated pattern's major structural features are the supinated feet, the relative outward rotation in the bones of the legs, and the relative tilt between the pelvis and chest creating a long straight back.

that our evolutionary heritage has given us and it is what makes us the unique and highly energy-efficient movers that we are.

Because of the properties of these two patterns we can describe them as both structural and functional phenomena and we can address them in our structural assessment and evaluate them in our tests.

These patterns create certain relationships between structures and in our structural assessment we can see the patterns being "held" to greater or lesser extent throughout the person's body.

As these patterns are the product of our ability to walk and run they are also possible to assess in function. We can take the altered relationships and we can put them into functional motion to see if they have the ability to move as we would expect.

By using the patterns to orientate us in the person's structural and functional relationships they become our entry point into the complexity. The benefit of using the patterns is that it gives us a simplified description of the person's current structural and functional status without losing the holistic perspective. We can use the patterns to zoom in on smaller and more narrow relationships and properties and keep the holistic connection to be able to zoom out again and explain how our intervention was connected into the entire entity. This is the best way to avoid stumbling into any reductionist pitfalls.

It is important to understand that we do not get stuck in the idea of the patterns, but acknowledge them as theoretical entry points and appreciate the ability they give to be able to zoom in and out in our explanatory models.

By using the patterns we also get a clear insight into where the person is today, in structure, in function, and thus also in ability to undertake development and be somewhere else tomorrow. The patterns become the starting blocks to the process-orientated journey that we want to take each client on. Read more on this subject in Chapter 14.

Our simple conclusion is that to study ensomatosy is to dedicate oneself to a never-ending journey of exploration and to learn to trust the intelligent body.

CHAPTER 12

The positive intention

Linus Johansson

Shoshin, "In the beginner's mind there are many possibilities, in the expert's mind there are few."
—Shunryu Suzuki

When working with a client or patient you are always faced with the interface of communication, interaction, engagement, touch, and movement. This interface is unique in so many ways and these profound aspects do not appear in any other circumstance than between two human beings. In this book the major focus is the actual structures and function of the human body. This is not a book about psychology or the human mind. However, we can never deny or exclude the immense importance the mind has and how closely connected it is to our structures and especially movement and function. In this chapter we acknowledge this and present a clear and simple method for working with a person's mind in relation to our perspective.

We should remember that we, our clients, our patients, and ourselves are all fellow human beings. This gives us a certain obligation to take some responsibility for their mental status and health, even if we have not been trained specifically to deal with psychological issues. Just being a fellow human being that listens and reflects on a person's thoughts and questions during a conversation can sometimes be enough. We don't have to have a degree in psychology to be able to do that much, just some decency and common sense.

However, as a therapist or movement teacher it is always the outcome of the intervention that is the major focus point. How can we affect and change the state that a person is in and take them to a more "optimal" version of themselves? For a therapist it is generally pain reduction that is the goal and for a movement teacher it is more often the progression and different

stages within the movement that is the focus point. Almost always there is a combination of the two, because movement and pain often go hand in hand, as discussed in Chapter 2.

To be successful and to get the desired outcome of an intervention will always depend on what happens between the two people in that situation and not just what manual technique or specific movement the practitioner chooses to use.

As you are aware, we are not machines with fuses that have burned out and need to be changed or buttons that have been turned to off and need to be flipped on again. Sometimes we might wish that we were that uncomplicated but (un)fortunately we are much more complex than that and it is all due to our highly-developed minds. We are cognitively advanced and highly psychologically driven beings. Just as we are sensitive to touch and movement, we are equally sensitive to what happens in verbal communication.

In this chapter we acknowledge the fact that the human mind plays a significant role in the outcome of our interventions. Therefore, instead of seeing it as a hindrance or obstacle we are instead going to invite you to the simplest of methods to make the human mind our greatest asset.

The way of the positive mind

What many of us have been taught in our training is to assess the body and its functions and search for injuries, difficulties, deficiencies, and reduced functions. We have also been taught the name of these findings and how to document them accordingly. The names given to the different pain patterns, held postures, or actual damaged tissue have long and complicated names, sometimes in Latin. The intention is to be able to do a standardized and proper documentation of the assessment and treatment, and to have a unifying language that we can use when communicating with another practitioner.

Most of the methods are designed to address and fit a certain problem that has been assessed and documented. This is all good because the intention is where it should be and the practitioner's focus is on helping the person with the issue presented. However, from our perspective, there is a major flaw in this approach. Let's elaborate on why.

Basically, when a client comes to meet us and get our help they are almost always in a state of pain or loss of function. Mentally they are rarely in a positive mode and their relation to their pain and body are not always the best. Right from square one the client is in a negative state of mind and has come to us to seek our help as they have lost the ability to solve the problem themselves. Their fate is put into our hands.

If we follow the old paradigm we will go down the same negative path as the client and we will start to look for all the "bad things" to seek out the problem. We assess and document what is painful, weak, tight, insufficient, and so on. All this is done with good intentions of course. The purpose is to address the issue and change it for the better using the very specific, narrow and reductionist methods that come with this paradigm. Our patients are in general with us on this approach. They know and understand how the current system works and they have probably experienced this many times before and firmly believe that "this is how you do it."

However, this is where we beg to differ. We do not believe "that this is how you do it." We believe that there is another way—a much simpler and, seen from our perspective, more powerful way.

If we summarize what we have so far it is the following:

1. **First** is the supposed cooperation between practitioner and client; a cooperation where both parties see, perceive and describe all aspects from a negative point of view.

2. **Second** is the language barrier. The practitioner describes and talks about the client's body using a language that the client does not necessarily have any insight into. When the client describes their pain or lack of function it is often done with everyday terms that are based on the client's life experiences, experiences that are difficult to standardize, making it very tricky for the practitioner to convert the information to fit into the given frame of the paradigm.

3. **Third**, there is the level of kinesthetic development. However this is when a small paradox steps into the picture. To be able to understand this we need to rewind a bit and take it from the beginning again.

A single joint or muscle can be treated and taken care of in many different ways: massage, manipulation, laser, shockwave, dry needling, and so on. All are recognized methods of treating pain. However, to be properly taken back to function we cannot just settle for local pain reduction. We also need to create development and reintegrate the person's entire movement system back into gravity. This integration will always start at the client's given level of kinesthetic development. Each will have a unique level of kinesthetic development and that level can differ a lot between different people.

If the client is a "mover" and uses his or her full movement potential the reintegration of the altered structure will be easy and require less from the practitioner. On the other hand, if the client is sedentary and not very well connected to movement the reintegration may not be as easy, or it may fail altogether, rendering any good work wasted.

However, it is not only the client's level of kinesthetic ability and development that plays a role in this arrangement. The practitioner's abilities are just as important. In many senses, the practitioner's kinesthetic awareness and abilities will be what the client has to rely on and develop from. From our perspective it is crucial that the practitioner knows how to engage and integrate motion in his or her own movement system to be able to develop the client's abilities.

"One must practice what one preaches, it's first when you move yourself that you can move others."
—Linus Johansson

Now, let us take these three aspects, the negative approach, the language barrier, and the level of kinesthetic development, and turn them into the most powerful tool in the toolbox.

The first of these, the negative approach, might be the simplest thing to change. What we always need when working with a client is a process-oriented mindset. To create that mindset you do not go down the negative path, you go up the positive path.

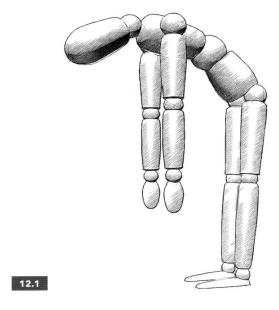

12.1

To illustrate the different approach we can use a classic example. A person bends forward, with straight legs, to try and to touch their toes (figure 12.1). The person stops halfway down due to "restrictions." The observer can express this finding in two very generalized ways, and with two quite different outcomes:

1. **The classic.** Wow, you are stiff! You can't even touch your toes! You need to work on your flexibility and stretch more. Why haven't you stretched? It's quite obvious why your back hurts!
2. **The positive.** You know what? I see a lot of potential for you to be more flexible. That's cool, and I know what we can do to try and fix that! And when we've made you more flexible you will probably not have as much pain in your back. Would you be interested in trying?

Both statements are tweaked to be each other's antagonists, but in fact both are not far from the truth, especially the first one. What we aim to show with

these two polarized examples is that they state the same fact, although expressed in two very different ways. Regardless of how the assessment is expressed the practitioner's intention is the same: to increase the range of motion for the client and make the movement easier to do and reduce the back pain. However, that is not the point we are trying to make: because what is interesting is not what the practitioner says, but what the client hears.

You might think they are the same thing, saying and hearing, but they are most definitely not. The practitioner has a clear intention of what to do and can more or less use any statement to express that without altering the set intention. However, what the client hears and certainly gets most affected by is whether the statement has a positive or negative tone to it. The content of the statement is secondary because it is not even certain that the client fully understood what the practitioner said. What is primary and what the client picks up is the tone of the message and in the two given statements there was a clear difference in tone.

The first statement, the negative, didn't add anything new to the picture for the client. It only told the client something they probably already knew and also were aware of when doing the test. This way to express an assessment, with a negative tone, also puts the practitioner in an even more authoritarian position. Once again, this can still work as the client thinks, "this is how you do it." However, we do not believe that a relationship ordered this way would ever be very fruitful.

The other statement turned the perspective. It gave a positive tone and suggested the idea that there was a start to a solution close at hand. Rather than making the client more aware of the limited range of motion, a positive statement like this instead leaves the person more curious and interested in themselves and in what can be done to increase the range of motion. Using a statement like this makes it much easier to get the clients process oriented and makes them more prone to engage in striving to work towards a solution.

The concept of the word "strive" is essential in this perspective. Creating a goal to reach is one thing; however, it is not when one reaches the goal that change happens in the body. It is while the person is striving toward that goal that the changes and development take place. This means that we can set proper goals for a client but gain the biggest development when striving to reach them. One could argue that this effort, in itself, is the goal.

All this taken together may also help the practitioner to be able to place less emphasis on his or her authoritarian position and engage at a more equal level with the client.

We therefore firmly believe that one should choose one's words wisely when working with a client. Choosing a positive approach with an appropriate language will lead to the client perceiving both you as a practitioner and the situation as positive and interesting, something that never can be wrong.

We would also argue that one becomes more present when one assumes a challenge with a positive mindset, and as Walt Whitman once put it "we convince by our presence." Our presence helps us to show even more that we are there for our clients and for no other reason. A good presence will also make them believe in us more, something that in turn will make them participate even more in becoming a part of the solution.

> "We convince by our presence."
> —Walt Whitman

What is more interesting is that creativity and ingenuity also go hand in hand with a positive mindset. A positive mindset will affect you as well as your client. You see and perceive things more clearly and solutions present themselves much more easily when your mind is positive. Inspiration comes when we choose to see things as possibilities and not as hindrances.

This way of setting one's mind when working with a client connects with the expression *shoshin*, used in Zen Buddhism and martial arts. It means "beginners mind" and refers to the openness, eagerness, and lack of preconceptions that a beginner would bring to a situation.

The quality of the content in the expression *shoshin* is profound when working with the human body and mind. To try and keep that openness, eagerness,

and lack of preconception can be difficult as a practitioner when the years start to pass. One gets comfortable and does what one has always done, and there is nothing particularly wrong with that. It can still help and be the base of a good treatment. However, it is still stagnation not to choose to evolve within your practice.

The major reason for choosing to stop is probably because it is quite hard work to keep pushing oneself to develop. However, having a positive mindset that always keeps you searching for the possibilities, potential, and abilities, be it in the human structure or any other aspect, will always keep you moving forward whether you intend it to or not. It is when we move forward and beyond that we can truly develop ourselves in a progression parallel to that of our clients.

Woven into the positive tone in a statement is the language it employs. As mentioned earlier, there is an academic vocabulary that therapists and practitioners are taught and that they use in their practice. The flaw with this language is that our clients may not have the same understanding of what we are saying as we do ourselves, and as we stated earlier, we believe that communication is a key element in a successful relationship. Misunderstandings or lack of understanding can quite quickly become counterproductive and diminish the chances of success.

The language barrier also contributes to that authoritarian state of mind that we do not want as it conflicts with the idea of a good relationship. What would be more suitable is if we could instead use a language and terms that are more anchored in the common language that both parties can relate to. This without losing the ability to document the findings and being able to report it to another practitioner.

In Chapter 5, we present such a language, based on movement and not pathology. Using a more common language like that will invite our clients to more easily understand and relate to the work we are doing with them. This, once again, will help us to create a good and fruitful relationship with our clients.

Let us move on from verbal language and instead look at the language of the body, also known as movement.

The ability to translate the spoken words into desired movement will be the next crucial step for anyone working from this perspective.

It is one thing to claim which movement needs to be reintegrated in a human structure and why. It is a completely different thing to actually manage to convert that intention into action and reintegrate it into a client's movement system, especially as we are referring to complex and dynamic movements through gravity and not just a leg curl in a machine or a contraction of some "deep core muscle" in a prone position.

We believe that to become a successful practitioner one needs to truly undertake the art of movement, just as much or even more than one would undertake the theoretical studies.

The fact stands that we are sprung from a lineage of movers, not thinkers. Our cognition has just been ours for the last fraction of the time that we have been moving through evolution. No one can dispute the fact that we are made to move and also most certainly made to learn through movement. Before we could learn things on a conscious level we learned everything subconsciously. The intriguing thing is that you have done that yourself. You once learned how to roll, crawl, stand, walk, and run all by yourself. No one taught you that, you taught yourself.

Therefore we advocate that you first sort out your own movement before you try to sort someone else's. This can be a harsh statement but try to explain to yourself how you can take your clients further if you cannot do what they need to be able to do. They rely on you to have cleared the path for them, to have put in the hours, to have built your awareness and love for movement.

We are not saying you should be able to run a marathon or be able to deadlift three times your own weight. What we are saying is that you must explore, investigate, enjoy, love, and foremost live movement to be able to do your job as a professional practitioner. You need to practice what you preach.

When you have first built your own awareness in movement, then you can help someone else do the same thing.

The method

When addressing, explaining, and interacting with our clients we always do it with a positive tone to set their mind on a positive journey toward a solution and to awake the interest in their own body and its potential.

When we also set ourselves on the same positive journey we don't see anything but possibilities, potentials, and abilities in our clients.

We create a process-oriented treatment plan based on our positive findings and by using a dynamic and understandable language we invite the client to participate.

We show our love and devotion to movement and we inspire our clients to build their awareness and to choose to develop.

Summary

If the human form and structure is a complex and a maze-like phenomenon, it is nothing in comparison to the human mind. However, the body and mind are not two separate phenomena; they are two indivisible aspects that make up the human being.

Our main interest in this book is human form and structure rather than the mind. Having said that, we want to underline that we never deny anything regarding the human being, but always acknowledge everything. What we have presented in this chapter is therefore our take on how we acknowledge the mind's influence on the human structure and how we can use this as a tool and asset to connect even deeper into the human form and function.

PART II

Introduction

In Part I we have given you an insight into our view of human form and function. We have stated our principles and the perspective we use to see, describe, visualize, and acknowledge human form and function. Our approach has been quite broad and not specific to any topic. We have given an account of our principles and not of any specific methods.

To let this first part crystallize into something more methodical we have written a few chapters to show you how we funnel the principles into methods. Please note that the following only are a small selection of the infinite versions of methods that this perspective can bring.

CHAPTER 13

A more detailed look at posterior tilt of the ribcage

Martin Lundgren

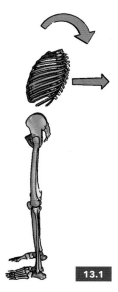

13.1

In this chapter, we are going to take a closer look at posterior tilt of the ribcage. In Chapter 8 we talked about the ribcage posteriorly tilting in relation to the pelvis; in that context there was a posterior bending throughout the spine, including the thoracic spine. But in this chapter we are talking about a position: a posterior tilt and a posterior shift of the ribcage. If we look at figure 13.1, there is an example of a posterior shift and a posterior tilt of the ribcage in relation to the pelvis. Again, this describes a static position and not a movement.

13.2

If we start to make things more detailed, we can divide the ribcage into a lower part and an upper part. We can then see that it is actually the lower part that follows this pattern of a posterior tilt and a posterior shift. In figure 13.2 you can see that the lower part of the ribcage has posteriorly tilted and posteriorly shifted in relation to the pelvis. If we look at what happens to the upper part of the ribcage, we can see that it actually does the opposite. In figure 13.3, the upper ribcage anteriorly shifts and anteriorly tilts in relation to the lower part (and the pelvis),

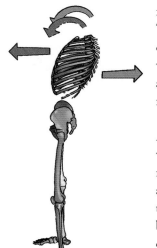

13.3

indicated by the blue arrows. This means that we get a compression in the ribcage, which is accompanied by an anterior bend in the spine, as in figure 13.4.

A bend is a series of tilts, which is here illustrated by the many arrows in front of the spine, every vertebra anteriorly tilts in relation to the one below (the arrows and the color of the vertebrae here are all the same color in the lower ribcage and the upper ribcage).

In this chapter, we are going to focus on what happens to the lower part of the ribcage. If we start at looking

13.4

at the vertebrae, it is my belief that some shifts are going on between the vertebrae. In figure 13.5, we can see that the vertebra above shifts posteriorly in relation to the one below. In this case, the vertebra in red is T12, and the one in green T9. This is the same as saying that the vertebra below shifts anteriorly in relation to the one above, as in figure 13.6. This series of shifts creates a kind of "false extension" in the body, meaning that we can have an anterior bend in the spine without making it very noticeable (if we did not have this posterior tilt/shift we would lean more forward).

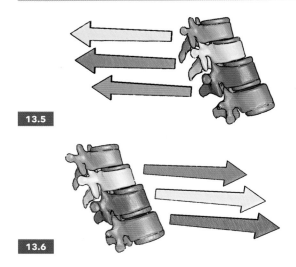

13.5

13.6

To know the exact position of a vertebra in relation to another is of course not the easiest thing in the world. If we only rely on our touch and our sight to assess the position of a vertebra, it is quite reasonable to question the validity of our claims (and mine). This is not to say that we cannot be correct using these methods, just

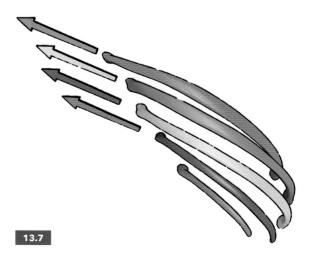

13.7

that there is a need for another more objective form of assessment if we want to make our claims more valid, like an X-ray, for example. Nevertheless, I think that assessing movement in this context has a higher degree of validity, especially if we use spring testing as described by Hesch.[1] In the case of this pattern that we are describing here, a spring test in an anterior and superior direction to a vertebra in relation to the one below is usually restricted. After different forms of interventions, the spring testing very commonly becomes less restricted (and the lower ribcage less posteriorly tilted; see figure 13.20 for an example of this).

If we continue and take a look at what happens to the ribs we can see that the posterior shift is translated into the ribs as well, with the rib sliding posteriorly in relation to the one below, like in figure 13.7 (rib 8, is also included here). In figure 13.8, the same thing is happening, but the vertebrae are included as well. When we look at this, we can see that the posterior tilt of the lower ribcage is in part created by these shifts between the ribs and vertebrae. If we have somebody lying supine, we can spring test a rib anteriorly to see how well it moves anteriorly in relation to another rib. Usually, we will find that the movement is somewhat restricted when we have a pattern like this.

Because the rib that is below is smaller and is anteriorly shifted in relation to the one above, it tends to "hide" in a superior shift in relation to the rib above, as in figure 13.9. Furthermore, because of the posterior shift of the ribs, the angle of the rib in the rib above tends to move medially in relation to the rib below, as in

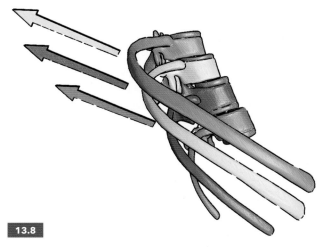

13.8

figure 13.10. If we feel where the angle of the rib is on one rib, and slide superiorly and feel where the angle of the rib is in the rib above, you can usually feel that the angle of the rib is more medial (and posterior) in relation to the one below.

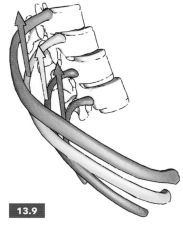

13.9

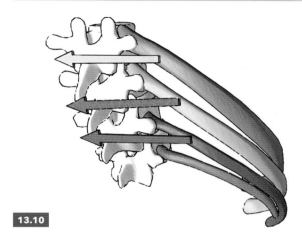

13.10

Moreover, we tend to get less space between the ribs, which sometimes even makes it harder to distinguish one rib from another; when you follow one rib you slide over to another one, thinking that it is the same rib, when in fact it is not. This is illustrated in figure 13.11. If we describe the interrelative position the ribs have to each other, we can see that the rib above is shifted closer to the rib below, as in figure 13.12 (in relation to a more "optimal position," described later). If we look at frontal relationships from a posterior view, we can see that same thing is happening, as in figure 13.13. We can also describe this as the rib above laterally tilts in relation to the one below.

This is interesting, because when we side bend away from the right side, we want the opposite to happen. If you put a finger between two ribs in the lower ribcage on the right side and tilt the ribcage toward the right you are probably going to feel that the ribs squeeze together, and when you tilt the ribcage the opposite way you are probably going to feel that the ribs separate from each other. This is illustrated in figure 13.14 where the rib above medially tilts in relation to the one

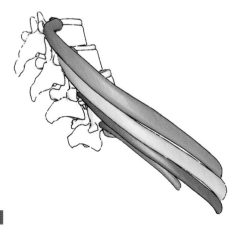

13.11

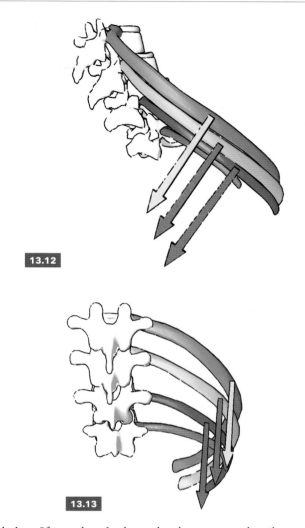

13.12

13.13

below. If we take a look at what happens to the ribs when we get proper posterior bending throughout the spine. The rib above slides anteriorly in relation to the one below, as in figure 13.15 (if we have a posterior shift of the vertebrae in relation to the vertebrae below). And if we look at what happens in the ribcage when we rotate we can see that the rib above rotates away from the one below, as in figure 13.16. All these movements mean that we get a shift away in the rib above in relation to the one below, as in figure 13.17.

All the above movements are related to the late resupination in the foot, i.e. in the foot on the same side, as described above. This is described in Chapter 8, in figures 8.50, 8.51, and 8.54. These movements happen at one side at a time, meaning that the movements are also related to rotations in the body. If we have one side where we lack these movements (the right, for example), that side is probably going to look more like figure 13.12. And if we have one side where

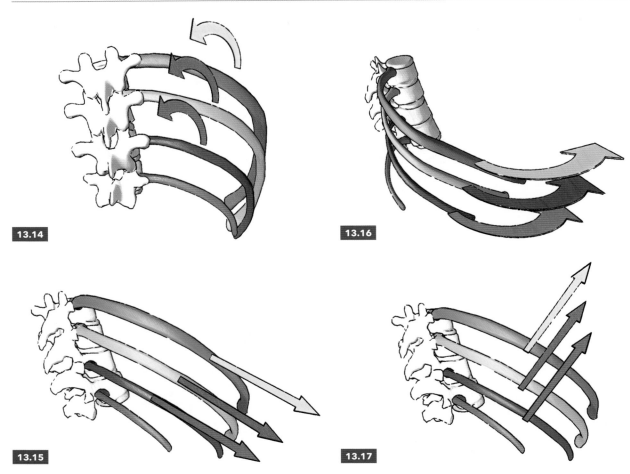

13.14

13.16

13.15

13.17

we have these movements (the left, for example), that side is probably going to look more like figure 13.17 with more space between the ribs (a position, not a movement). If that is the case, we are probably going to have a ribcage that is rotated to the right in relation to the pelvis, and we are probably also going to have a harder time rotating (and tilting) the ribcage toward the left in relation to the pelvis.

The earlier resupination when the pelvis is posteriorly tilting in relation to the ribcage is also relevant in this context. The lift in front of the lumbar region and posterior shift of the lumbar region in relation to the ribcage and the pelvis can help out with creating less posterior tilt and shift of the lower ribcage (illustrated in figures 8.57 and 8.58 in Chapter 8).

Another aspect that can be interesting to explore is the ribs' relation to the vertebrae and the pelvis. If, for example, we tilt the ribcage toward the left in relation to the pelvis (and the pelvis tilts toward the right in relation to the ribcage). The ribs then tilt

toward the right or inferiorly shift in relation to the vertebrae, but lift or tilt toward the left in relation to the pelvis, as in figure 13.18. If we tilt toward the

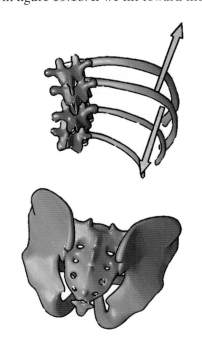

13.18

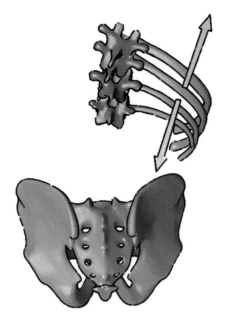

13.19

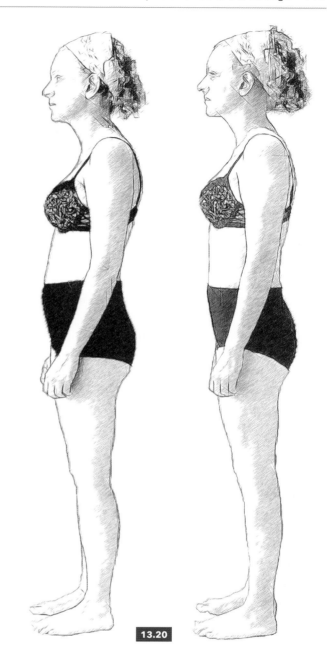

13.20

opposite side, as in figure 13.19, we can see that the opposite happens. The ribs superiorly shift in relation to the vertebrae but inferiorly shift in relation to the pelvis. The ribs' ability to superiorly shift in relation to the vertebrae can then affect the ribcage's ability to tilt, toward the right in this case. If you put your hand between the ilium and the lower ribs on the right side and grab the lower ribs and carefully push them superiorly and tilt the ribcage toward the right, and hold it there for 1–2 minutes, you are probably going to find that it is easier to tilt toward the right afterward. Sometimes even the ribs on the left become freer after this. (Be careful with the ribs and know your limits in how much you can press.)

In figure 13.20 you can see an example of a session that is done with creating strategies and interventions mainly from the principles we have described above. In the after picture to the right, you can see that we have less posterior tilt and shift of the lower ribcage, and we have more separation between the ribs as described earlier in figure 13.17. The work that was done here was mainly done with working directly with the bones, which means working distinctly with a bone and its relationship to another bone, one rib to another for example, or one vertebra to another (or a rib to a vertebra). There was also work done with the sacrum and its relation to the ilia, following the same pattern of resupination (not described in this context).

Note

1.   Hesch (2015).

Reference

Hesch J; The Hesch Method; integrating the body, recognizing and treating inter-linked whole-body patterns of joint and dense connective tissue, and reflex dysfunction. Workbook. Aurora, CO: Hesch Institute, 2015.

CHAPTER 14

My method

Linus Johansson

"As to methods there may be a million and then some, but principles are few. The man who grasps principles can successfully select his own methods. The man who tries methods, ignoring principles, is sure to have trouble."
—Harrington Emerson

The words of Harrington Emerson could not be more true to me. They are also the essence of what we try to capture and present in this book. In this chapter I will give you my personal perspective. I will compile what you have read up to this point and turn the principles into "my method." Please note that this chapter is an expression of my own views and does not necessarily reflect the views of the other authors of this book. The purpose of this chapter is to describe how I work in relation to what you have read up to this point.

My first reflection as a practitioner is that it is all too easy to become method-orientated in this line of work. I believe that we have lost touch with the basic idea of principles and that we are at the moment drowning in a sea of methods that sometimes seem completely unattached to any principles or even sound reasoning.

Much of the argumentation and reasoning regarding human function are based on methods rather than principles. All too often practitioners see the "problems" in a person's body and apply a generic method to solve them. I rarely see someone appreciate the entire person as a system of potential and by using sound and solid principles, take on the job to develop the entire person.

Methods have somehow become the principles and we are happy with the argument that "movement is always

good and it becomes better the more complex it is." This generic recommendation, in my opinion, is about as useful as a fork when you are served soup.

One reason why "methods as principles" is so common in this line of work today is probably due to the digital era we are currently in. Methods have become extremely easy to show and distribute. You no longer have to do the great work of writing a book to show the world what you "know" or "can." All you have to do is start a blog, a video channel or some other platform in the social media stream. With your mobile phone you can now distribute whatever you like. The thresholds are very low and self-criticism is often lacking.

When you look out into the social media stream and you see people in the fitness, movement, or therapeutic realm, presenting pictures and video clips, what do you see? Is it principles or methods? Of course you see methods! The reason for this is that you can very easily capture a method in a 30-second video clip or a series of pictures. This is something that is impossible to do with a principle. A principle needs a book like this to reach an audience, rather than a short clip online. It is not hard to see which one of them is the easiest to consume.

I also believe that, with this never-ending stream of methods, we have only created more confusion for our clients. How can a person figure out what to do when there are millions of options and there is no distinction as to who should do them or why? How can that be inspiring?

I firmly stand by my opinion that we need to bring forward highly-trained therapists, movers, and trainers that are orientated by the principles they believe in. Unlike a method, a principle is something you learn over time to understand and believe in, something that you carry with you in your profession. A principle also lets you seek guidance and inspiration when needed. A well-grounded principle will always be a solid stepping stone when you want to develop yourself and others.

A true professional will therefore always let the principle be the algorithm that, together with all the variables each client provides, suggests the methods needed to create successful development for that client. The method is never constant, only the principle.

I would be the first to admit that this can be somewhat complex and challenging. I was once one of those who thought that if I only got this next gadget or learned these new exercises it would help me resolve all my clients' needs. My conclusion now, looking back, is that each new method was yet another corner to round, just to find a new endless straight road of options with no distinction or consensus. Getting people to move using generic methods is one thing, and not very complicated.

However, understanding both how and why a specific person should be able to move to develop their abilities is something quite different.

Therefore, paradoxically, it was when I started to move backward, instead of forward, that I first actually managed to reach onward. Everything started to become clearer and somehow more understandable when I put aside the idea of chasing methods and instead started to search for principles.

In my clinic

When people ask me what I do I don't say that I am a physiotherapist because that is a title that is laden with too many preconceptions. That is also an answer to what I am, not what I do. The answer to the question of what I do is that I work with movement integration. Movement integration to me is to see, appreciate, and develop a body through movement in relation to the body's internal relationships and to gravity.

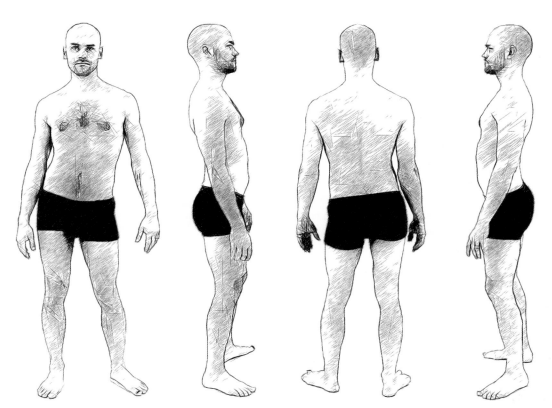

14.1

The structural relationships within a patient viewed from all sides.

When a person comes to see me at my clinic we have always several sessions booked ahead. I never meet a person only once, as that would be unfair both to them and to me. I need to know that I can see them continuously for several sessions to be able to have a successful treatment plan.

When a person seeks out my help they always have a question regarding themselves, a subjective experience of their body's function and a story to go along with it. My first job is to sort this all out and get an overview of the first variables they present.

To do this I always see a person as two bodies: the conscious body and the subconscious body. The conscious body is the person's mind, wish, will, and ego, and that part of the body that is intentionally controlled, in other words the skeletal muscles. The conscious part of the person is also the part of the person that I interview using words.

The subconscious body represents everything else: in other words, everything that a person does not have any will to control or that they are not consciously aware of. I address the subconscious part of the person during the structural assessment and the movement tests.

During the first session I take the medical history to get a timeline of what has happened up to this point and what they wish to get from our encounter. Regarding medical history, everything is potentially of interest and can give us important pointers into the future and what needs to be done. The problem is that a client can very easily get stuck in the past. They try to explain and blame past events for the problems they have now. However, until someone invents a time machine we can't really do anything about the past. The only thing that is truly interesting is therefore where we are today and where we need to be tomorrow. What was yesterday is history and we can't change that. What we can change is the future and that can only be done if we are present today. My intention is therefore to encourage them to let go of what has been and to create a glimpse into a not too distant future where they can be in a body with different abilities and hopefully less pain.

From here I also sort out how they go about during an ordinary day at work, or engaging in hobbies, leisure

activities, and so on. I try to map out as much of this as possible to find out where I can most easily get the opportunity to gain the biggest leverage when it comes to reintegrating movement into the system.

People always have different relationships to movement. Some are more interested in it, some are less so. If a person already has an interest in movement, that is always a positive starting point. If they do not the job will be to get them interested.

The great thing is, however, that it doesn't matter what movement they do. Thanks to the fact that all human movements, and our principles, are based on the gait, this will give us direct access to be able to work with and improve all human movements.

This then becomes the greatest of all "tricks"—to get each person more involved and interested in themselves. Letting them understand how I believe that their movement can be improved will hopefully make them more committed to the interventions I present. A high degree of commitment will always lead to development, both structurally and intellectually.

In the structural assessment the client stands in gravity and I assess how they situate themselves through

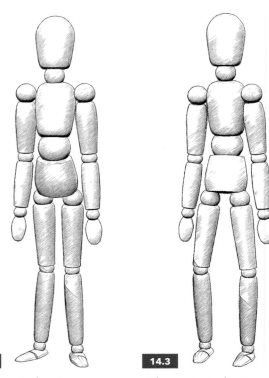

14.2

The pronated pattern.

14.3

The supinated pattern.

their internal relationships and how they relate themselves as an entity in relation to gravity. I use the two patterns, "pronated" and "supinated" (described in Chapter 11), to orientate my interest in their structure and function.

From the visual assessment I formulate a suspicion that there might be potential for development in certain relationships between structures. In order to confirm or reject my suspicions I do movement tests for the relationships to see if movement is present or not and in what degree and of what quality. I base these tests on the biomechanical rhythm of the gait. The tests are made in big movement patterns and all the way down to small local relationships.

From this I get both the specific results from the test and also the bigger picture of how connected the conscious and subconscious bodies are, i.e. on what level their body awareness is.

All these variables then go into the algorithm that my principles constitute. From this I obtain a fully personalized set of methods that, in the most effective way and time, will create development and hopefully fulfill the person's wishes.

To summarize: the principles I use as my algorithm are:

+ The body is constructed to be as energy-efficient as possible and it will always strive for least resistance.
+ We are forged to stand and walk on our two feet, i.e. the gait.
+ The gait is built up out of a wave of biomechanical relations in the human form based on the connection to gravity.
+ All other movements that we can do are based on the gait.
+ A person is always more prone to engage if communication is based on a positive and process-oriented language.

I then keep looping through these principles during the entire treatment plan, always reassessing the outcome of my intervention and reevaluating the results I get; this is necessary because when methods are applied, the body will alter and transform, giving new variables and therefore new input into the algorithm and hence new methods to work with.

When starting a treatment plan I have marked out the relationships of interest in a person's body and how they are all connected in the biomechanical rhythm of the gait. My intention is then to restore the structural and functional relationships to be able to strive toward the idea of the person's "optimal body," as mentioned in Chapter 5.

The process can begin at any given point in the body. Although it might look as if I am working with one "part" of the body my intention is always ahead of my intervention and resides within the idea to alter the relationships within the entity.

The pure work I do is both manual techniques on the treatment table, manual techniques in motion through gravity, and pure movements done by the person themselves.

When I work with a client I always acknowledge that what the person's movements contain, and what my hands engage in too, is nothing less than every single tissue and structure in that person's body, from the deepest neurons to the most superficial epidermis and everything in-between. Be it bone, blood vessels, nerves, lymphatic glands, adipose tissue, interstitium, fascia, muscles, the soul or what have you, everything is there for a reason and is a part of the vast and complex entity that is the human form. To ignore this or think anything else would, to me, be to completely reject the idea of the indivisible body.

I therefore do not assess, judge, or speak of the tissue as single structures and I do not give them individual qualities, such as tight, weak, overactive, soft, hard and so on. To me they are all one and cannot be divided in to separate phenomena. The only time you can successfully divide tissue is when using a scalpel and to be able to do that in practice you need dead tissue. My view on this is therefore that if you divide the living and moving human form, in theory you also, theoretically, kill the tissue. You diminish the tissue by dividing it into made-up, constructed, and titled parts. Just as it is impossible to put a dissected body together again, to me, it is also very difficult to get back

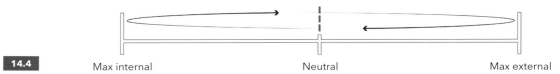

14.4   Max internal   Neutral   Max external

This scale represents the maximum range of motion for the internal and external rotation of the femur in relation to the ilium. In this example the structure is held in a relative neutral position on the movement scale and has the ability to exert maximum internal and external rotation.

the picture of the indivisible human form if you first theoretically pull it to bits. I therefore firmly believe that with the right principle and the right interventions to go along with them one can still keep the idea of the indivisible body and work with it as an undivided unit. My recommendation is therefore that you put down your mental scalpels and try to see, appreciate, and enjoy the human form for what it is, a truly indivisible entity. In other words, ensomatosy.

When I describe the findings I perceive in a person's movement apparatus I focus on what to me are more constructive parameters that I can truly work with and develop. First, I always use the two movement patterns as my foundation, both in structure and function. From that I then describe the potential I see in the different relationships and how they are affecting each other. In each relationship of interest I then describe the dynamics of the present movement.

The dynamics of a movement is a more holistic way of describing movement. In the description of dynamic both the actual movement that is possible is described and the range it can increase. However, lack of movement is also described as it is a potential object of development.

The term "dynamics" derives from the idea that there is always an optimal range of motion for a relationship; this range of motion is theoretically described from a neutral position on a movement scale, as in figure 14.4. What happens in real life is the fact that the structure is not always in neutral position; however, the subjectively expected range of motion is still the same.

What happens subjectively when a structure is not in neutral, as in figure 14.5, is that the person describes the feeling as stiffness because they cannot exert the full range of the expected movement. In this example the "neutral" position of the hip is relatively more internally rotated. The experienced stiffness firstly depends on the fact that the person cannot exert more internal rotation between the femur and ilium as it is already, more or less, fully internally rotated. A person often expresses this as, "It feels like bone against bone." Secondly, they also lack full external rotation, as there is no more length in the tissue to reach that far over the movement scale to the end of the external rotation. A person often expresses this as, "Look! I can't go any further, I'm too stiff."

By seeing and describing the dynamics of a relationship in this way I can connect them further on into the person's movement chain and see how they depend and relate to all other relationships. This gives me a systemic approach to develop a person's abilities, from the most local movement all the way out to their total engagement through gravity.

By using movement patterns and structural relationships connection and the dynamics of movement I have a powerful methodology to incorporate the complexity of human function without getting lost in the details. It also gives me a more vivid picture of the person's potential and a very good

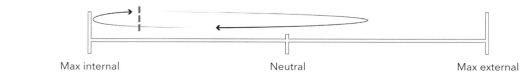

14.5   Max internal   Neutral   Max external

In this example the structure is held in a relatively more internally rotated position which shifts the person's "neutral" position on the range of motion scale to a more unfavorable position.

platform to be creative and inventive when it comes to finding solutions for my clients' needs and wishes.

In conclusion

Remember the words of Ralph Waldo Emerson,

> *"The man who grasps principles can successfully select his own methods. The man who tries methods, ignoring principles, is sure to have trouble."*

Be sure to first sort out your own principles. What do you believe in? Which are the principles that you follow and that will guide you to the methods you need to do a good job? Find them and hold on to them, and let them develop your clients and you.

Should you ever find that the principles you believed in have failed you, and did not fulfill your expectations, do not hesitate to search for better ones. Remember, because you believed in the first principles and they were a part of your development thus far, you can now differentiate and seek out new principles to engage in and to work with.

Also remember that nothing is ever constant: you are always in a state of transition in your development, whether you want to be or not. You just need to realize that and enjoy the ride.

I have found my principles and I am in a very clear state of transition where the principles truly are developing me. This fantastic fact makes every patient a new and unique challenge and each time I use my principles the algorithm they construct grows stronger and more complex, helping me to help others. I develop myself by developing others—it is a great feeling.

I can honestly say that this is what makes this way of working so incredibly creative and satisfying. Knowing that each day will present new challenges for me is what gets me up in the morning—that and the kids.

CHAPTER 15

Soma Move®

Linus Johansson

"My heart is pounding hard and the sweat is running down my face, yet my breath is not out of control. I breathe in harmony with every movement I make and I can feel myself smiling.
Each fiber in my very being is alive and active.
I move over the floor in what seems to be a never-ending flow of movement.
I close my eyes and I feel completely focused, I am here, I am now and I am movement."

Many years ago I started my journey to take my bachelors degree in physiotherapy. My curiosity about the human body and its function led me to choose physiotherapy as a profession. One of my goals with my chosen education was to get answers about how the human body truly functioned.

After three years of hard studying and finally taking my degree, it dawned on me that I hadn't received the answers to my questions. The next realization came rather quickly when I left the safe environment of the university and got thrown into the world of hard and uncontrolled reality. I quickly understood that, instead of being all-knowing, which I had hoped for when starting my education, I knew nothing.

The problem was not working as a physiotherapist. I had been given very well thought-out methods that fitted the well-established paradigm, and performing my duty as a therapist was no problem. The problem was that when I questioned

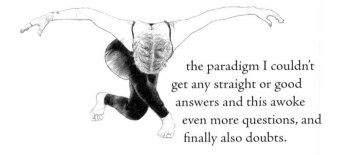

the paradigm I couldn't get any straight or good answers and this awoke even more questions, and finally also doubts.

After many years of pondering, searching, reading, and studying I came to a new realization. The simple reason my questions never got answered during my education was that there were no answers to give. The paradigm I was currently active within was not created to answer those questions. It was created for something very different.

This realization in itself became the answer to me and it set me off on another journey. That journey brought me to the conclusion that there is no single, unifying and absolute answer to the question of how we function—and there never will be. But instead of being struck down by this realization, I now understood that I could find my own answer to my questions, just as everyone before me had done.

To me, this book is a part of my quest. I see this book as one of many milestones on my endeavor. I know that I will never find the answer to my question, and what I also know is that the answer to my question is not what I am truly after. It is the constant striving toward the answer that gives me the opportunity to learn something new every day. Being able to learn something new each day for the rest of my life is now my answer, which in other words means that I am living in the very center of my goal, I have already reached it. Each day I reach it.

One part of this journey is to explore movement. Movement was also the base of the first steps I took when I started my exploration and I haven't stopped yet; I am still exploring movement everyday.

We speak a lot about the complexity of the human structure and function in this book and we try to give you our perspective on it. What is interesting, and we will be the first to admit this, is that when we put all of the complexity together it will always end up as movement. When you read Gary Ward's chapter (23), you will also see that he has come to the very same conclusion, as have many others before us.

This makes movement the all-covering concept that always will hold all of the human complexity, even if we undertake it or not. The complexity will always be there when we move, all the different tissues in the body will get engaged, all the different systems that adapt and control will fire, and all our thoughts and feelings that make us undertake movement will continue to conduct us along the way.

Now there is quite a difference between movement and movement, and that is one of the reasons we have written this book. We wanted to illustrate our vision of what movement is and therefore also invited the contributing authors to give you a more vivid and broad insight into this. The culture and ideas surrounding human movement are endless and one's standpoint must therefore first be clarified and expressed for others to be able to create a relationship to it and have a discussion around it.

When I personally speak of movement, I always refer to an all-engaging, holistic movement that makes a person flow and float through gravity with ease and grace. In my opinion movement is supposed to be all-engaging, energy-efficient, beautiful to behold, and always perceived as a pleasant effort for the person executing it.

The movement I practice goes hand in hand with the principles and explanatory models presented in this book, which perhaps is quite obvious, but what is intriguing is that my movement and these principles have always been in conjunction, even before I started to try and shape the principles into words and sentences, together with Martin. The reason is simple: the complexity and the principles behind human movement have always been there and always will be there, whether we choose to explore them or not.

Soma Move

It all started as a hunch, a feeling that there must be something more than this. At the time I was stuck in the gym, doing the classic exercises trying to gain muscles and slim myself in to something that society told me was the desired look for a man. I struggled with this for many years, ending up with the feeling I was running in a hamster wheel, struggling without getting anywhere.

At that time I was a physiotherapist working in the very heart of the fitness business in Stockholm and I was very much influenced by that environment. After a couple of years we moved north, bought a house in the countryside and I started my own business, something that let me loose and gave me the chance to rethink and to explore. I left the "old" movements and began exploring the "new" movements. (Please note that old and new refer to my own knowledge and understanding, and not a generic timeline.)

Instead of being, like before, influenced by others around me or from the social flow, I tried to find my own way. Playing, feeling, enjoying, and exploring my own potential. Before I was more or less dependent on equipment of all kinds to be able to perform a proper workout: kettlebells, barbells, dumbbells, rings, benches, sticks, ropes, and so on. What I tried to do now was to just devote myself to my own body and the interaction it had with gravity and the ground beneath my feet.

After many years of exploring I had constructed a movement concept that I was very keen to share with others. The opportunity was given to me during a fitness trip abroad. I was one of many presenters on that trip and had the chance to let a group of people experience my concept during our stay. In this group was Cecilia, my now very dear friend and colleague and one of the contributors to this book. She was also a presenter on this trip and we took part in each other's concepts during these two weeks. We soon became fascinated by the fact that we presented and expressed movement in a very similar way, although we had never met before and, most intriguingly, came from two very different backgrounds.

Cecilia is a yogi and dancer, and she bases almost all her inspiration from a gut feeling. She listens to what her body tells her is the "right way" to move, and she follows. I, on the other hand, was forged in the old school of physiotherapy and taught that everything I did had to be anchored in literature and scientific articles.

Our coming from two very different places and still expressing movement the same way was the main reason why we couldn't let this opportunity slip past us. From that moment we started to work together to explore and develop what has become the movement concept SOMA MOVE®.

Soma is the Greek word for "body" and put together with "move" it became our playful interpretation of what "human movement" is and what it could be.

My intention, when I present a Soma Move session to a group of people, is to give each one the opportunity to "meet themselves in motion." To me Soma Move is all about using a curious mind to explore one's movement potential.

"Because you explore your body's potential today, it will reward you tomorrow."

The foundation of Soma Move

Soma Move is based on some fundamental evolutionary conditions and interpretations of qualities that reside within the human form. Underlined and presented within the framework of this movement concept we believe that we have created a platform where you can go back to basics and touch on the very foundation on which all your movement qualities reside.

I will in this chapter present my interpretation of the basis of Soma Move. In Cecilia's chapter (20), she presents her interpretations of movement, with the four qualities: grace, rhythm, power, and stillness. These four are also important parts of the foundation and fused together with my principles they create Soma Move. When you read her chapter you will find where the resemblance between us came from and yet how different we are. Our similarities and differences are the strength of our relationship.

Flow

The first and most important focus point in Soma Move is to move in a constant flow, in uninterrupted movements performed at each person's individual level. The purpose of this is to come close to the same kind of meditative state that you will get when you are walking or running, the two fundamental movements that are the pinnacle of a flow. We never stop, never hold, never hesitate, we just move in a continuous flow of motion.

The variation of movements is much greater in Soma Move than in walking and running; however, each movement is like an island in the steady stream of flowing motion. On each of these islands of movements, the opportunity to not think and just do is presented. Exactly as when you walk and run, you never think, you just do and go with the flow.

The goal, however, is not to make big, expansive and over-exaggerated moves that look like a circus performance. Each person must explore their own body's potential and understand where they are today in capacity and ability. However, regardless of what level a person is at, she or he can always move in flowing and continuous movements. The aim is flowing from one position to another, being strong, fast, and controlled and at the same time soft, slow, and free.

The breath

Needless to say, the breath is not a new tool for anyone working in the holistic realm of movement. For many hundreds of years practitioners of holistic movement have understood the power of the breath. In the fitness industry, however, knowledge of the breath has not yet reached a higher understanding than that it is used to "oxygenate the blood" and "somehow stabilize the core" whilst squatting or deadlifting—I am well aware that I am pushing it a bit here, but you also know that this isn't too far from the truth.

In Soma Move we establish the importance of the breath in every session we present and let our participants indulge in the breath's richness and complexity without forcing it onto them on an intellectual level; we would rather have them experiencing it on a kinesthetic level and reaching the understanding that way. It is also the breath, in harmony with the person's movements that assures the flow of motion, focus, and awareness we pursue in every Soma Move session.

The term "harmony" is paramount to us. We encourage our participants to explore the idea of breathing in harmony with every movement they perform. Giving them the feeling and insight that the breath is just as important a movement in the body as any other motion or exercise. Turning each session into a journey of exploration from every perspective.

The concept of the expression "harmony between breath and movement" is based on the following three key elements:

First element

When breathing, we always ask our participants to take each breath in through the nose and out through the mouth. Passing the air in through the nose will make it cleaner, warmer, moister, and more turbulent than breathing in through the mouth, hence making it more suitable for the lungs.

The exhalation is done via the mouth due to the fact that it gives us the opportunity to control the flow of air leaving the body. Breathing out through the nose can only provide one constant resistance and no possibility to modulate the flow. With the tongue and lips we can control the flow of air and in that way create a resistance that makes us engage even more whilst moving, especially where and when more control and awareness is required.

Second element

We see the breath as built up out of four stages. The first stage is the inhalation, through the nose, toward filling the lungs to the desired level; the second is reaching the top of the inhalation before turning it to a controlled exhalation via the mouth. The third stage is the exhalation toward emptying the lungs to the desired level before ending up at stage four, where the breath is turned, once again, into an inhalation via stage one. We

loop through these four stages continuously throughout our entire life.

To match the four stages of the breath with the movements performed and to create the desired harmony we illustrate each movement as a compound of positions and journeys.

In each movement performed in a Soma Move session we say that we reach a "position" and between each position we say that we perform a journey toward the next position. Please note that these positions are not static or held positions. They are merely a springy turning point where we load the tissue through the principles of elastic recoil to set us off on our next journey. The harmony between the movement and the breath is therefore very important and with these positions and journeys we can seemingly weave in the four stages of the breath in every movement we do.

Third element

We also see the inhalation and exhalation as two fundamental movements that "close" and "open" the body.

When saying "close," we see it more as "the flesh" holding or grasping on to the skeleton and generating support and control rather than creating an actual folding

or flexing movement. This, in other words, does not mean that one cannot breathe in during long and tall movement, in fact rather the contrary. Creating a closed connection to one's body during an inhalation is a very good way to create elongation, length, and height in any movement.

The "closing" happens during stage one and especially stage two. The respiratory system is connected, both directly and indirectly, to all moving parts in the human form. During the inhalation the respiratory system engages in a contractive fashion to create a negative pressure in the lungs. This will in turn send a wave of engagement out into the rest of the movement system to assist and complete the desired inhalation. It is this engagement that we see, appreciate, and use.

Our interpretation therefore tells us that the feeling of wanting more support and control in a movement should be the guide to when we inhale in a certain part of a motion. To anyone doing heavy squats or deadlifts this is not news.

With "open" we refer to the ease that the exhalation creates and the ability it gives to let go and to create space and length in the system. This openness happens during stage three and most clearly during stage four. Exhaling and letting go of tension in the respiratory system will set off a wave of ease into the rest of the movement system, creating the opportunity for tissue to lengthen eccentrically and reach the springy end point of the journey.

We acknowledge that other movement cultures may see the interaction between movement and the breath differently. We therefore underline that this is our interpretation of this unique interaction and that through this system we can explain many events and find great progress in our way of moving in relation to the breath.

Hands and feet

The hands and feet have always been our connection and movement interface to the world around us. As a child you explored your surroundings and developed your relation to gravity using only your hands and feet. It has always been that way and will hopefully remain so.

In this simple movement we illustrate the breathing pattern in a movement. In each position to the side we breathe out and on the journey in-between we breath in.

In Soma Move we therefore pay tribute to our evolutionary heritage that is still the foundation of what we are today and we make the hands and the feet, once again, the most important "parts" of the human form when in movement.

During a Soma Move session we move and carry ourselves through gravity, using only our hands and feet to suspend us over the ground in a constant flow, never weight bearing or even touching the ground with any other areas of the body (apart from the knees on occasion). Letting the feeling of the active engagement spread from the hands and feet, up into the arms and legs, into the shoulders and hips to meet and to unite, effortlessly, in the center of our body and movement. Creating connection, consciousness and continuity.

Moving in this way and with this intention also gives the clear sense of moving as one indivisible unit.

All aspects, the movement, breath, hands, and feet, give support to each other and to the structure as a whole.

Length

When we move in a Soma Move session we always strive for length. However, striving for length is actually more striving for organization and not necessarily an actual measurable elongation. That means that a movement or position can be fully flexed or compressed and still contain a lot of length.

Unlike the phrase "posture," which is filled with preconceptions, the term "length" is easy to comprehend and easy to learn how to use. Length, unlike posture, can also be created in many different places, such as an arm, a leg, a hand, a foot, a hip, and the spine or even in the breath.

To lengthen becomes very much a holistic concept and it contains a lot of complexity without being complicated. By lengthening an arm and focusing on reaching further with your hand and fingertips you will get an engagement through the entire arm, into the shoulder, and further down into the body, connecting the length with some other intention and helping to organize the body to perform the next perfect and efficient movement.

The sensation length gives, in combination with a flowing and a centered movement, is profound.

The fact that the concept of length is easy to grasp, to apply on many different structures in the body, and to integrate in movement makes it a very powerful concept when participating in a Soma Move session.

Elastic recoil

If we take all these aspects (flow, breath, length, and the engagement in hands and feet), together with the four qualities that Cecilia presents (grace, rhythm, power, and stillness), and put them into movement we can get a very intriguing engagement in the body. When you move this way you allow and ready your body to engage in the most perfect, well balanced, and foremost, energy-efficient way that you can.

As we discussed earlier in this book, moving in the most energy-efficient way is the very foundation of our evolutionary heritage. Through hundreds of thousands of years our bodies have evolved to become perfect movers and those unique properties reside within us all. The marvelous fact is that we all have the ability to access that heritage if we please.

There are a lot of aspects that can be honed to contribute to increase energy efficiency and they are all deeply intertwined and dependent on each other. One of these aspects, and the one we hold highest in our Soma Move practice, is the ability to keep the flow in our movement. Each hesitation and break in the flow costs a lot of energy, both of body and mind. That is why it is important to learn how to listen with the body and to take in new movement via our kinesthetic abilities rather than using the conscious mind. Our bodies are far more intelligent than we could ever be.

"The important thing to think about is not to think too much."

Reaching a flow in motion is almost like entering a meditative state. The mind can relax and the body can take over. It is a blessing to be able to let the intelligent body be in command and give up all thoughts and ideas that can interrupt the flow.

Another aspect that is highly energy inefficient is compensatory movement patterns. When one is in pain or has lost an ability to move in a certain way the body will compensate to fulfill the mind's wishes. We spoke of this earlier in Chapter 2. One can say that pain from wear and tear of the body is inevitable; however, one can also say that it is possible to work to avoid the compensatory patterns that pain can start. The way to do that is to explore. In general, the ego dictates and the body follows; with our "exploration" we instead turn it around and let the body dictate and the ego follow. However, it doesn't have to be pain that turns the tables, just the pure feeling that "my body doesn't want to do this, it wants to do this instead!" That is all that is needed. To use movement as the scope to perceive your body's wishes is probably the biggest realization one can ever have.

That is why we, in all Soma Move practice, invite our participants to explore every movement to find out in what relation the body wants to do this. Then, and only then, can we truly connect inward, let go of all our preconceived ideas, and just be one with the flow.

Some might argue that it is necessary to punish the body and push it to be able to develop and reach the desired goals. In Soma Move that is not the case. We believe that development comes from showing respect and love to one's body. To care for and praise the potential that we all have residing within ourselves is the only thing that will create development. The goal is just to be in movement, something that means that we reach our goal in every Soma Move session we do.

Moving further along the lines of aspects that contribute to energy efficiency is the ability to load tissues in an eccentric fashion, i.e. to create elastic recoil. The old school tells us that muscles contract, and they do, but not before they have elongated through the interaction with gravity. This is important to grasp as it is a keystone to all efficient human movement.

When we move through gravity and suspend our body over the ground, using only hands and feet, all tissues will load eccentrically in movement and elongate before they can recoil and take us further in our movement. It is inevitable: our bodies are forged to function this way and it is the only way we can create truly efficient movement.

Simplified we can say that all the complex structures of the body, e.g. skin, fat, muscles, fascia, bones, nerves, blood and blood vessels, etc., are engaged in elastic recoil, in one way or another. What is important to understand is that we can never hold one tissue responsible for the function of elastic recoil, for they are all in it together.

If we continue to keep it simple we can also say that elastic recoil has almost the same properties as a rubber band. The complex structures of the body will create a recoil when elongated eccentrically with a sufficient speed and load, and thereby give a much richer quality to the contraction and the approximation of structures bound between the recoiling tissue.

This richer quality is most easily reached when we compile all the earlier mentioned aspects, power, length, grace, breath, rhythm, engagement, stillness, and flow, into one movement. If we then also add the intention to perform this compiled movement as quietly as possible then we are very close to the perfect motion.

This can be visualized most easily by imagining a big cat stalking its prey, moving without making a sound. It can prowl, charge, and land completely silently and its body is moving in perfect flow and harmony. Those are the qualities that awake when you move with the intention to move absolutely silently.

You can try this very easily. Stand carefully on a chair and jump down onto the floor as you normally would. You will make quite a loud noise. Notice also how your engagement in your structure is when you are doing this.

Now alter your intention. Get up on that same chair, still with care, and now jump down on the floor and do everything in your power to do it completely silently.

Did you notice how your entire body altered with that intention? You became more aware, softer, still strong and controlled. You became the cat.

We want to have the exact same intention when we move in a Soma Move session. This intention lets us heighten awareness in each movement and create a flow of movements that are soft and controlled and at the same time powerful and vivid.

Centering

In conclusion, what we aim to create in a Soma Move session is to give the participant the feeling of a pleasant, interesting, and constant energy-efficient flow of movement. A flow where it seems that everything is connected, from the hands to the feet, from the breath to the mind, and from intention to action.

All this will of course happen to various degrees and is fully dependent on the participants' abilities and earlier experiences. However, what we always aim to give each person is the experience of being centered, and what the feeling of being centered is can only be answered when you participate in a Soma Move session. Therefore, we welcome you to meet yourself in movement.

CHAPTER 16

A note on craniofacial development

Martin Lundgren

This chapter takes a brief look at craniofacial development and how it relates to movement. Orthodontist Mike Mew has suggested a syndrome which he calls craniofacial dystrophy.[1] This syndrome involves a downswing of the whole "anterior craniofacial structure" (ACS). Mouth breathing is one thing that is mentioned as a cause for this downswing. Several studies have found that breathing through the mouth can cause changes in the development of the cranium.[2] Mouth breathers tend to develop a more convex facial profile and a longer face with a narrower maxillary arch with a higher palatal vault.[3] Mouth breathing is even associated with growth retardation in children.[4] Furthermore, having the mouth open with the jaw dropped makes it impossible to keep the tongue in contact with the palate, which is considered crucial for our craniofacial development.

Mastication is also another thing that is recognized as an element affecting the developing face.[5] Craniofacial dystrophy seems to be a modern phenomenon and people further back in history who ate a denser food appear not to have had the same issues with crowded teeth and a downswing of the ACS.[6]

This downswing is highly relevant to our movement potential as it can hinder our ability to extend properly, which will be described later. The tongue is the upper end of the Deep Front Line[7] (figure 16.1), which means that there is a continuity from below. It should, therefore, be reasonable to think that what happens below can affect the tongue's ability to expand the palate and push the maxilla up and forward in relation to the rest of the cranium. At the same time, what happens to the cranium and the maxilla will also determine if the tongue has something to work against. One study

found that total nasal obstruction resulted in an extended head position in all cases.[8] So again, if you cannot breathe through your nose, it is impossible for the tongue to keep contact with the palate, not only because the mouth is open but also because the head is posteriorly tilted. If this posteriorly tilted position is chronic it might hinder our craniofacial development, which again can make it harder for us to breathe through the nose. Looking at it from this point of view, it is reasonable to think that the body as a whole and its relationship to gravity, i.e. a person's overall kinesthetic development, is an important element in craniofacial development.

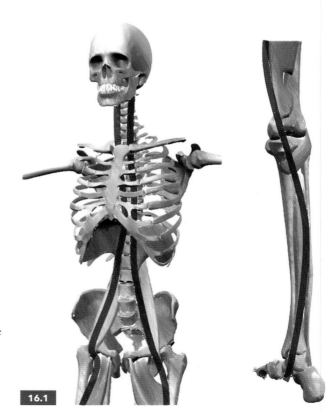

16.1

We will now take a closer look at what this downswing entails when it comes to movement, and explain it in terms of relative movements. In Chapter 8 we looked at how the pelvis and ribcage interact. We can extend this relationship to include the head as well. When we get posterior bending throughout the spine and the ribcage posteriorly tilts in relation to the pelvis, as in figures 8.54 and 8.55, the head should be able to anteriorly tilt in relation to the ribcage, as in figure 16.2. The position described earlier, the extended position of the head with the nasal obstruction, would mean that we anteriorly shift and posteriorly tilt the head in relation to the ribcage, as in figure 16.3 (enhanced for visual purposes). If we are stuck in this position, we get less pressure from the tongue pushing the maxilla forward and up. In figure 16.4, we have divided the skull into ACS (yellow, without the jaw) and the rest of cranium (blue).[9] If the ACS swings down, it anteriorly tilts in relation to the rest of cranium, represented with the blue arrow in figure 16.5.

Now to the point. If we are in this position, with the ACS anteriorly tilted, it is not possible to get the same amount of anterior tilt of the head in relation to the ribcage when the ribcage is posteriorly tilting or when we are getting extension throughout the spine.

Because of the relationship between the ACS and the rest of the skull, the ACS can anteriorly tilt more in relation to the ribcage than the rest of the skull. For the rest of the skull to able to anteriorly tilt in relation to the ribcage as much as the ACS, the relationship between the ACS and the rest of the skull must change. The ACS needs to posteriorly tilt in relation to the rest of the skull, i.e. an upswing of the ACS. If we compare figure 16.2 with figure 16.6 we can see that the arrows are much smaller, we have a smaller green arrow above the skull, indicating that it is harder to anteriorly tilt the head in relation to the ribcage. This lack of movement is because the ACS cannot go any further, as the throat is in the way. In figure 16.7 the ACS is in the exact same position, but the rest of the skull has anteriorly tilted in relation to both the ACS and the ribcage. As indicated by this picture, this also means that it is easier to get more lift in front of the spine and get more extension throughout the spine. This anterior tilt is of course not a movement per se but a development that hopefully can happen over time.

It is my belief that manual intervention can be a great help and, in some cases, a necessary part of working with craniofacial development. In figure 16.8, you can see a picture of me before and after I started implementing manual interventions in line with the understanding and background mentioned above. The manual

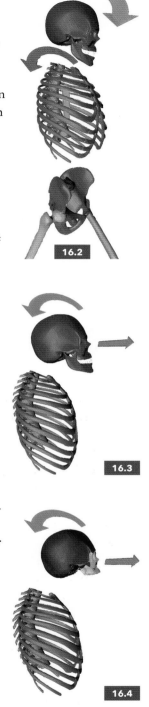

16.2

16.3

16.4

interventions include working directly with the bone, bone-to-bone relationships as well as with the soft tissues. The pictures are taken about a year apart. The main focus has been to achieve a proper tongue position, posterior tilting of the ACS in relation to the rest of the skull, and widening of the ACS transversally. I have found it very difficult to keep the posterior part of the tongue in contact with the palate without the aid of manual intervention. A lot of attention has gone into manually widening and lengthening the palate and the maxilla. A crucial element of this widening seems to be intranasal work. The relation between the occiput and the mastoid also seems like a critical area. I have worked with the hypotheses of anteriorly tilting the temporal bone and the mastoid in relation to the occiput and the zygomatic bone and the rest of the cranium, at the same time as posteriorly tilting the ACS (upswing), which seems to make it easier for the tongue to keep contact as well as freeing up the area around the occiput and the atlas.

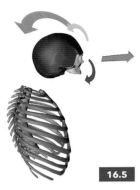

16.5

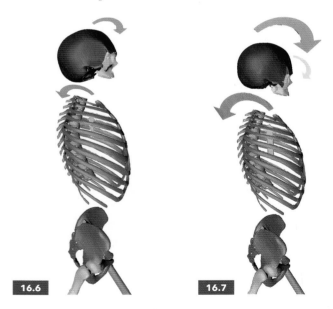

16.6   16.7

When I examined the superior aspect of an 80-year-old lady's coronal suture in dissection, it appeared to be totally fused. A thing to remember though is that all the inferior fissures between the occipital bone, temporal bone, and sphenoid bone (petro-occipital,

spheno-petrosal, spheno-occipital), are not considered to be sutures but to be synchondroses, i.e. cartilaginous joints, which would indicate that they are more predisposed to change or perhaps could be a place for growth even later in life.

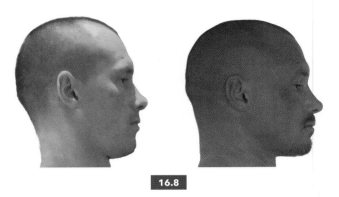

16.8

In figure 16.8, you can clearly see that the posterior part of the cranium (the blue part) has anteriorly tilted in relation to the ACS. If you look at the external occipital protuberance on the right-hand side, you can see that it is substantially higher up at the same time as the ACS has stayed in the same place. The two pictures are about the same height, but if you look at the width anterior to posterior, you can see that the right side picture is longer.

One hindering factor in the process seems to be the fact that I have extracted both lower third molars. When I bite together the mandible wants to slightly slide/shift anterior in relation to the maxilla because of the shape of the teeth. This seems to be a limiting factor when trying to achieve an upswing or posterior tilt of the ACS. This anterior slide/shift also appears to be related to the width of the mandible transversally, as the anterior shift seems to diminish momentarily when I have been working on widening the mandible.

It now looks like I have slightly more space where the third molars used to be. There is also more space in-between the teeth. I have always needed to struggle a bit when getting the dental floss between the teeth, which is not the case anymore. I never had any trouble breathing through the nose, but there seems to be more space intranasally than before, making it easier to breathe through the nose. It also feels like I have more space for my tongue and I need considerably less effort to keep the tongue in contact with the palate.

Notes

1. Mew (2014). Mike Mew has built on the work of his father (John Mew), known as orthotropics.
2. Harvold et al. (1981) and Basheer et al. (2014).
3. Basheer et al. (2014).
4. Morais-Almeida et al. (2019).
5. This is mentioned in Mew (2014). See also, for example Katsaros (2001) and Kiliaridis (1995).
6. Rose and Roblee (2009).
7. As described by Myers (2014).
8. Vig et al. (1980).
9. This division is undefined when it comes to which bones are included in which part. (You could make a case for a division between the visceral and the neurocranium.) But this undefined division should be enough to explain the overall pattern of the downswing I am aiming for here.

References

Basheer B, Hegde KS, Bhat SS, Umar D, Baroudi K; Influence of mouth breathing on the dentofacial growth of children: a cephalometric study. *Journal of International Oral Health* 2014, 6(6):50–55.

Harvold EP, Tomer BS, Vargervik K, Chierici G; Primate experiments on oral respiration. *American Journal of Orthodontics* 1981, 79(4):359–372.

Katsaros C; Masticatory muscle function and transverse dentofacial growth. *Swedish Dental Journal Supplement* 2001, (151):1–47.

Kiliaridis S; Masticatory muscle influence on craniofacial growth. *Acta Odontologica Scandinavica* 1995, 53(3):196–202.

Mew M; Craniofacial dystrophy, a possible syndrome? *British Dental Journal* 2014, 216(10).

Morais-Almeida M, Wandalsen GF, Solé D; Growth and mouth breathers. *Jornal de pediatria* 2019, 95(Suppl 1): 66–71. doi: 10.1016/j.jped.2018.11.005.

Myers TW; *Anatomy Trains: Myofascial Meridians for Manual and Movement Therapists, 3rd edn.* Edinburgh: Churchill Livingstone, 2014.

Rose JC, Roblee RD; Origins of dental crowding and malocclusions: an anthropological perspective. *Compendium of Continuing Education in Dentistry* 2009, 30(5):292–300.

Vig PS, Showfety KJ, Phillips C; Experimental manipulation of head posture. *American Journal of Orthodontics* 1980, 77(3):258–268.

PART III

Introduction

The purpose of this book is to give you an opportunity to see and understand our perspective on the human form and its function. We show you our interpretations and elaborate in our discussions on what we believe movement to be and how it is manifested in the body. This book is a milestone on our own journey; however, we would not be where we are today if it had not been for all those who cleared the path before us and those who walk beside us expanding this perspective. To show our respect and gratitude we have welcomed the opportunity to invite several of our colleagues and teachers to contribute to this book.

Every contributing author has had a great impact on our view of human form and function. They are all great contributors to the holistic realm of movement and therapy.

We asked the contributing authors to elaborate on the term "movement" from their own perspective and show how it is present in their life and practices. We did this because we wanted to show you the great variety of interpretations you will get when you ask highly trained and experienced practitioners in the holistic realm to contribute with their thoughts.

Our intention is, once again, to show and underline that there isn't just one perspective or one truth to the mystery of human form and function. Depending on our perspective and the interpretations we make, we might end up with quite different stories to tell.

We have very much enjoyed reading what our contributing authors have written and we are very proud to be able to present these beautiful texts to you in this book. We do hope you enjoy them just as much as we did.

Gary Carter

Gary Carter has been a myofascial manual therapist, personal trainer, shiatsu and craniosacral practitioner, anatomy and fascial anatomy lecturer, and also a yoga teacher for over 30 years. Gary has run and owned a movement and therapy center in Brighton for 25 years, Natural Bodies (naturalbodies.co.uk), creating a method of training, moving, and health, blending a lifetime's work of martial arts study, bodybuilding training, athletics, qi gong, and yoga. He has run yoga teacher trainings and numerous myofascial anatomy courses in the UK and Europe.

Gary worked alongside Thomas Myers in the late 1990s, teaching and introducing the Myofascial Meridians theory to the UK.

He has participated in many dissection studies over the past 15 years and co-teaches on some dissection courses for his students with Julian Baker. Gary has explored and created different myofascial continuums in the human body that are applied to human function, which can blend seamlessly with Soma Move.

CHAPTER 17

Variations in myofascial slings and continuities

Gary Carter

"Life requires movement."
—Aristotle

I was walking and chatting with my father recently in the street to shop for some food, we rounded a corner and I stepped over a corner step that was an entrance to a café. My father hadn't seen it. It was a step of about 15 cm in height. His foot hit the step, he tripped and started falling, but with quick reflexes he gathered himself by stumbling and running, regaining his position swiftly as a fit, agile 25-year-old might do. He looked at the step and exclaimed, "Well, that's just plain dangerous, some elderly person could have a nasty accident." My father is 81 years old! At this point, I said to him, "Do you realize that you would be considered 'elderly', but you just moved out of that fall as easily as a younger person would."

He is a Masters athlete, was European 200 meter champion at 76 years of age and again at 80, World champion 300 meter hurdles champion and 200 meter hurdles champion at 76 and 81.

I really understood in that moment that maintaining or even regaining full easy range of motion throughout our bodies, throughout our lives, is

Putting the hip and leg through a fuller range of motion regularly keeps the joint structures fluid and mobile; it drives fluid flow, drives lymph flow, allows the soft tissues to have a pliable range, and maintains tonicity in the tissues. The gluteal region, for instance, is able to be fully involved in all motions so a reactiveness is available in side-stepping, moving rapidly and efficiently around objects.

"… this ninety-five-year-old man came hiking twenty-five miles over the mountain. Know why he could do it? Because no one ever told him he couldn't. No one ever told him he oughta be off dying somewhere in an old age home. You live up to your own expectations …"
—Christopher McDougall, *Born to Run: The Hidden Tribe, The Ultra-Runners, and the Greatest Race the World Has Never Seen*

essential for our health and survival. To be able to live fully with our moving bodies allows us to live fully in our environment. That stumble my father athletically recovered from, for a regular "older" person could have resulted in a leg, hip, wrist, or arm break, or even a head injury.

Maybe we could consider "Movement as a life force."

Being able to stand from sitting, to squat and to stand from a squatting position is a fundamental motion that proves a struggle for many westerners and Europeans of all ages. A colleague and fellow movement teacher friend of mine works every year on reservations of the Native Americans (First Nations). In a conversation with a few elders, they had noted that the younger generations had trouble keeping up with the elders whilst running. One of the elders had stated that it was because the younger generations sat on chairs and not on the ground or the floor. "It changes how their legs function," he had exclaimed.

"All movement is life."
—Buddha

"Excess sitting is now linked with 35 diseases and conditions, including obesity, hypertension, back pain, cancer, cardiovascular disease and depression," says Dr. James Levine, director of Obesity Solutions at Mayo Clinic in Arizona and Arizona State University. "Governments such as Australia, Canada and the United Kingdom have identified sedentary life as a catastrophe." He goes on to say, "The current generations of children are likely to die earlier than their parents, many of the projected deaths could be due to diseases linked to sedentary lifestyles, Westerners now sit on average for up to 15 hours a day." We can see why this could be considered a catastrophe.

"Movement is our birthright."
—James Oschman (Lecture at Westminster University, 1999)

My experience is as a teacher of movement, a personal trainer, a manual therapist in shiatsu, craniosacral, and KMI structural integration practitioner as well as a yoga teacher and a myofascial anatomy course leader with ongoing research and studies in human anatomy and fascial function through dissection in many laboratories over 13 years, alongside Julian Baker, Todd Garcia, and Gil Hedley. I was also a founding member and dissector for the Fascial Net Plastination Project at the Laboratories of Gunter Von Hagens' Body Worlds. I have found that through dissective research I have gained a deepened view of understanding the "textures" of the body, and the incredible landscape of shape and form that an anatomy book cannot clearly give.

My studies in yoga, the teachings of my original teachers, and my movement work from the late 1980s were all influenced by the work of Vanda Scaravelli. She had a specific and unique way of moving within the context of yoga that was informed by t'ai chi, martial arts, music, and breathing practices, helping to create freedom and lightness in all movements. Vanda Scaravelli had a saying, "Going with the body and not against it." This absolutely resonated with me during my first explorations in the dissection lab. Shapes galore, meaning multiple potentials for movement and efficiency of movement; this had a huge impact on my movement, training, manual therapy methods, and teaching life, and also on how I saw the body.

To "go with the body and not against it" has many meanings and can help in achieving efficiency of movement and less potential for injury. Educating individuals in "connecting the dots" in movement, sport, martial arts, yoga, and dance, allowing them to find the "wholeness" and integration in motion is a fundamental element of my training methods, yet sometimes we may need to look at regions of the body to understand its behavior, exploring areas of the myofascial whole. This also has implications for how we describe movement through teaching, instruction, and treatments. For

example, using language such as "the pectoralis region of the myofascial body" is suggested when describing how a tone or texture change in this one region will influence the rest of the structure.

Myofascial connections have been an interest of mine and a way of working from my early days of study in structural integration, eventually co-teaching with and assisting Thomas Myers (author of *Anatomy Trains*) in 1998 to 2003, to my ongoing practices of exploring myofascial continuities in motion since that time.

Dissection changes both how we see and what we see, and looking at the body through a lens of fascial connections opens up a wide variety of possibilities, including different ways to explore and understand movement. As Julian Baker has said, *"We are not just dissecting bodies here, we are dissecting our knowledge, we are dissecting our language and we are dissecting our beliefs."*

This is a statement that is abundantly clear every time we go into dissective research. What also became apparent to me was a series of myofascial structures as continuities that are different to some maps already proposed and make sense of movement patterns in terms of their ability to be economical in effort. This has then led me to a field of research in movement, dissection and manual therapy for the past 15 years of a 30-year career.

The ideas presented here are from two of a possible nine variations in myofascial sling-like continuities from my forthcoming book, *Anatomy Slings, Variations in Myofascial Continuities*. I will also be clear that these are models created by the scalpel, remembering that any process of dissection is a process of creation by destruction—we are taking apart the whole and taking apart the integrity of the structure. These explorations are influenced by the myofascial continuities and will also continue to respect the 'whole' in this process.

We will look at the effect these short, sling-like myofascial structures have on motion by bringing an amount of attention to these regions of the body in various practices in movement and manual therapies with interesting effect.

So first a little background.

A new way of seeing

We are living in a time where the ideas of anatomy have been challenged with new views of our structure, different considerations of biomechanics, the reviewing of how our structure generates its stability and what constitutes "posture," whatever that may mean. We have had many visionaries create wonderful insights and new explorations of our anatomy and function which shift perspective, leading to creative and different ways of understanding manual therapies, training systems, movement practices, through to some medical procedures.

The study and research of the fascial system, in particular, has led to an explosion of new ideas as well as confirmation of what was already understood. The existence of fascia has long been known, with mapped illustrations of the structures in the 1800s and the writings of A. T. Still more than a century ago.

Here I explore a continued vision of our connected anatomy with yet another way to see and experience how some myofascial structures are connected/related to each other, by standing on the shoulders of the many creative thinkers who dared to put their ideas out there for further study and research. When we see something from a different perspective our whole landscape and strategies can change, hopefully in a more energy-efficient manner.

We will start by looking at various individuals who have helped shape our visions of what we think is "set in stone." These are just some key individuals along a particular route—there are many more that could and will be added in the future.

Andreas Vesalius was one of the first anatomists to gradually change the view of anatomy—how we see anatomy—revolutionizing the study, realizing that students needed to be involved in the dissections, claiming that Galen's textbook of anatomy was flawed and required replacing. He gave us images that showed anatomy of our structure in different ways.

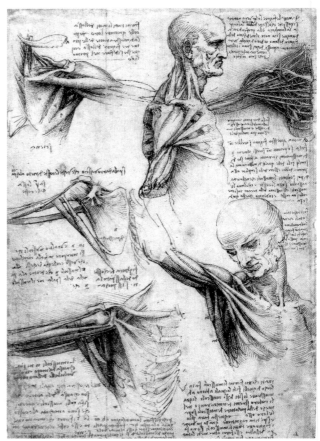

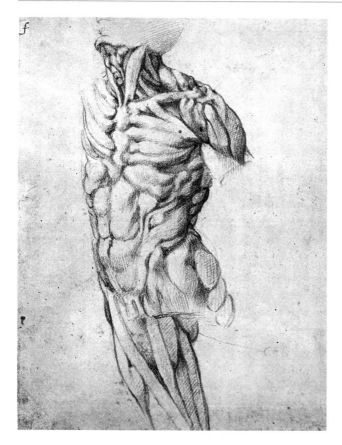

Along with Vesalius we had other artists and visionaries, including Leonardo da Vinci, an artist, designer, engineer, architect, with illustrations that explored the anatomy of form and workings out on paper of how things might connect ... to an inspiring illustration that suggests a 15th-century tensegrity idea of the human form.

Michelangelo, another artist, sculptor, architect and visionary, created exquisite workings to understand the

layout of form to such a degree that his statue of David had people swoon at its unveiling, such was its beauty and symmetry.

Raymond Dart explored myofascial continuities in the form of interconnected slings looping around the body. He graduated from medical school in 1917 and was professor of anatomy in Johannesburg in 1923. He was well-recognized for his anthropological investigations. Dart had a session of Alexander Technique in the late 1940s and maintained that it influenced his work for the rest of his life. Raymond Dart brought awareness to the double spiral arrangement of the human musculature, which connects the right side of the pelvis to the shoulder and to the neck, diagonally around the body in sheets of connected muscular structure, where we see the beginnings of a functional myofascial arrangement.

Professor John Hull Grundy was another visionary that explored layout and function in ways unseen in terms of anatomical exploration. Professor Grundy studied at King's College, London. In his early career, before joining the Royal Army Medical College, he lectured in art, and was also a lecturer in anatomy and architecture. He encouraged students to do their own learning rather than being fed facts verbatim.

Kurt Tittel was a visionary and pioneer and along with Raymond Dart explored a series of sling-type arrangements in the body, which we find in his seminal work *Muscle Slings in Sport*, first published in 1956. Tittel obtained his Doctorate of Medicine at the University of Leipzig and in 1952 he established a sports medicine practice. He was interested in

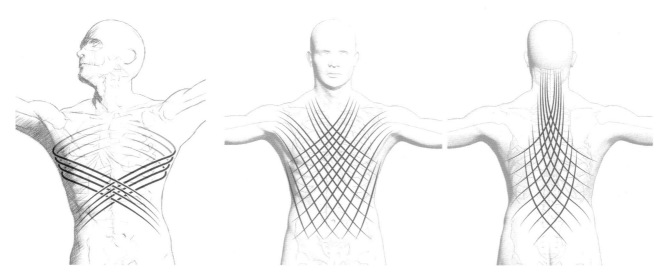

anatomy and more specifically, the functional anatomy of the athlete, and in 1964 became director of the Independent Institute for Sports Medicine in Leipzig, continuing to become the president of the Society of Sports Medicine. With over 500 publications in the field of sports medicine he is considered the father of sports medicine in Germany.

Tittel was influenced by the German anatomist Herman Hoepte, who shaped his viewpoint in exploring form and function. Tittel understood that muscle, bone, and tendon could not be fundamentally understood without considering their relationship to the whole individual.

Thomas Myers is a visionary, a practitioner of Rolfing and structural integration and was also an anatomy teacher to the Rolf Institute in the early 1990s. He developed a way of teaching myofascial anatomy in the method of a game, where he would connect up muscle along the front, back, sides, etc. of the body. Myers noticed that all anatomy books suggested a "single muscle" theory and his teacher Ida Rolf was suggesting they are all connected through the fascia. He was also shown an article by Raymond Dart linking the muscles through the trunk as a double spiral arrangement. Tom Myers's response to this was "why stop there?" and he extended the double spiral arrangement into the legs to make it a whole body structure, becoming the world recognized Spiral Line from the Anatomy Trains.

Myers then continued creating those continuities into the many connected muscle and fascial lines from head

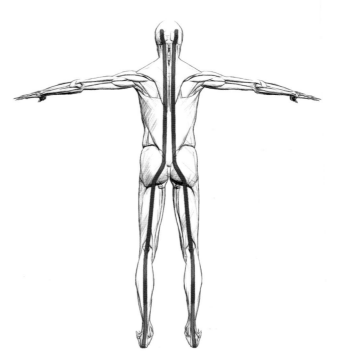

to toe. From this notion he produced his seminal work, *Anatomy Trains*, creating a connected whole body view of our myofascial body with a new way of seeing the muscle map as an interconnected network.

This once again is another change in seeing our structure and looking beyond the single muscle theory. This work has also been supported by the skillful directions from Todd Garcia, an anatomist and dissection tutor in the USA, under the watchful eye of Tom Myers.

It was Gunther von Hagens who developed plastination, a process that allows us to see the human body in different arrangements in a fixed 3-D form.

Plastination

Plastination, invented in 1977 by scientist and anatomist Gunther von Hagens, is the groundbreaking method of halting decomposition and preserving anatomical specimens for scientific study and medical education. A revolutionary method of cadaver tissue preservation, plastination permanently infuses the tissue with silicone rubber, enabling it to be viewed, touched, and studied with a level of detail and durability that was previously impossible. This method

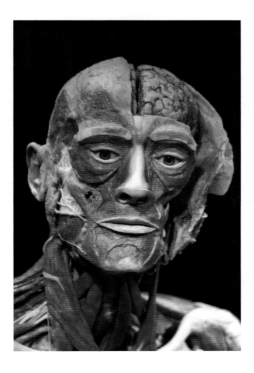

for displaying human anatomy for study and education grew to world fame and recognition through the Body Worlds international touring exhibitions, which helped to bring fascial anatomy education to the masses and yet again opened our eyes to the many ways we can unravel the human form.

The first plastinated fascial body

Exploring the plastination process has been the focus of a key group of people from the Fascial Research Society. In January 2018, fascia research scientist Robert Schleip, clinical anatomist John Sharkey, professor of anatomy Carla Stecco, and director of anatomy and plastination Dr. Vladimir Chereminskiy came together in the Fascial Net Plastination Project (FNPP), in collaboration with the Plastinarium in Guben, Germany, the world headquarters for the internationally acclaimed Body Worlds exhibition.

A collaboration of the Fascia Research Society, Somatics Academy, and Gubener Plastinate (GmbH), the FNPP is an international group of anatomists, bodyworkers, movement educators, acupuncturists, physical therapists, academics, professors, and physicians trained in fascial dissective methods. Together, they aim to advance human fascia anatomy education through the creation of the world's first 3-D human fascial net plastinated forms.

Similar to the Body Worlds exhibits, the mission of the FNPP is to create permanently-fixed human fascia specimens to help people understand the complex ubiquity of fascia, what it looks like, and where it is found. These specimens were exhibited for the first time at the Fifth Fascia Research Congress in Berlin, Germany, in November 2018. The exhibition marked the completion of the first phase of an ambitious, three-year plan to create the world's first all-fascia, whole-body plastinates that will become a part of Body Worlds.

Carla Stecco, MD, orthopedic surgeon and professor of human anatomy and movement sciences at the University of Padova, Italy, said, "This is a unique opportunity to get outside the laboratory and communicate the meaning of fascia in a direct way to all

people—to show its continuity through the body, its resilience, and its perfect structure. This project will open the eyes of any who, until now, were not confident about the importance of fascia."

All these creative pioneers have managed to expand our vision and allow our eyes to see beyond the ends of an individual muscular structure into a more "whole"-istic, unified world.

A spring-loaded system

Look at the behavior of any animal, and you will see a familiar pattern: rest, movement, pandiculation, easy relaxed movement, alertness, food, chase it, eat, rest, stretch/pandiculate, saunter, run, trot, climb (if it can), jump, rest, be alert, adrenaline of being chased, drink, breathing hard, breathing slow, and so on. I have just described our cat on a regular sunny day, chased by the dog, climbing, sauntering, eating, drinking, chasing a mouse, fighting back at the dog, sleeping. Animals possess all the components that facilitate a balanced pattern of life and sustaining its "myofasculature," articulatory network; keeping the springiness in the system is essential for ease of motion, lightness, grace and poise.

If I look at my father's regular movement patterns while training for Masters athletics, which is performed with varied durations, it is not too dissimilar to the descriptions above. There are long easy runs, easy lengthening work, plyometric style movements, mobility work, stretching and loosening work, sprint challenges, plyometrics, weight training with light and heavy loads, and then competition, the adrenaline, the release of energy, breathing hard, breathing soft, increased fluid flow, eat to meet the appropriate demands, rest, lightness, ease of motion (which becomes important for efficiency in sport). A similar sequence will assist in a balanced pattern of everyday moving life.

When engaged, the muscle tissue tensions the fascial surround, drawing it in closer; this tensioning increases the intramuscular pressure, which continues it to assist the force transmission throughout the structure, ideally maintaining efficiency of motion.

All fascial tissues store amounts of elastic and kinetic energy and these vary all over the body. Our tendons, ligaments (to a different degree), and fascial tissues all have the potential to store and release elastic energy.

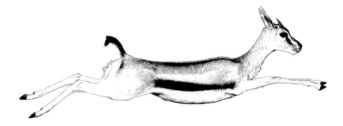

Given the fact that we humans have a lot of elastic materials in our makeup, the tissue can displace and rebound. With the fullness of fluid in the system this occurs to a greater degree.

It is known that deer tendons, and also horse and camel tendons and ligaments, have been used to make catapults, bow strings, and cordage from the Roman to the Saxon era and beyond, due to their tremendous elastic strength.

Kangaroos display interesting elastic potential through hopping. Research found a spring-like motion in the kangaroo's legs. The fascial tissues, including the tendons, are tensioned as an elastic band would be; the consideration here is that the tendons are acting as springs and muscles are tension generators. This enables an efficient release of stored energy which gives rise to the enormous jumps kangaroos are capable of without a huge demand on oxygen consumption. Gazelles display similar properties.

Scientists have since found that the mechanisms of force transfer between muscles and fascial tissues in humans have a similar kinetic storage potential to that of the kangaroo and gazelle during walking, jumping, and running; the stiffened stored energy makes this possible.

If we drop a stone to the ground the structure does not rebound much or, depending on the material, it will shatter. Drop a rubber ball and its elastic qualities allow it to bounce many times. If we drop a stainless steel ball onto a hard surface, its initial bounce is quite high—in fact it will bounce higher than a rubber ball as the stored energy is greater in the stainless steel as it returns its energy quicker. The key here is the stored energy.

"Motion is lotion."
—Eric Dalton

Because our tissues are primarily elastic storage mechanisms, our ability to move in different directions, rebound, absorb and disperse forces is part of our nature, as it is for any other creature that is myofascially oriented. This quality in our tissues can diminish with the aging process, less movement, illness, pain, and injury. Lack of motion can stiffen tissues, fascial planes and gliding interfaces between muscles can densify with each other, limiting movement.

The elastic fiber organization can become less organized, adding to a decreased range of motion. Muscle, when not used, can go through the process of atrophy. Think of the arm in a plaster cast; after even just two weeks of immobility it becomes visibly smaller. As muscle tissue diminishes more collagen is laid down. Collagen is a stiffer material (Stecco) so makes movement a little more restrictive.

As we age we start to lose muscle mass, and again collagen increases, so people will notice the stiffening as aging.

"It's all a matter of slowing down the loss of muscle mass as we age."
—A. Carter, World and European Masters Athletics Champion

It is a process of finding tensionally balanced movements around the system, in load bearing, body weight, and resistance work.

It is known that the fascial tissues are responsive to temperature, so as we stiffen a little through collagen laydown, colder temperatures will affect this more (Antonio Stecco, Ulm University). Maintaining a degree of balanced muscle mass has a metabolic process which helps to maintain a warmer environment in those tissues.

Tensionality

"There is no up and down in the Universe,
only in and out."
—Buckminster Fuller

The counter-forces mentioned in Fuller's quote are both aspects of the same thing, existing only in relation to each other. Compression elements create tension and vice-versa. When exploring various forces occurring through our myofascial/osteofascial body we can look at it in terms of tensile forces and compression forces all dancing together at any moment in time.

The simple act of going upstairs, turning to pick something up, leaning over to grab a coffee cup whilst holding something, the other hand will employ a myriad of these combined forces. Simply standing or walking is a balance of forces with freely useable energy throughout our system. However, if imbalanced it will create more effort and load on various regions of the body causing possible strain, restricted movement, and even pain.

Model skeletons are usually depicted standing on a frame in the osteopath or physio clinic, a hospital university classroom, Pilates or yoga studios. There is a pole put through it from below and clips, wire, and screws attaching it at the joints. Take these props away and we find it on the floor in pieces, as skeletons appear in archaeological digs.

If we could magically disappear the attachments and film the skeleton falling, we would notice that it

would probably spiral to the ground. This occurs because all joint endings are rounded, curved one way or another, from shallow to deep curves.

"The infinite perfection of connections."
—Buckminster Fuller

Tensegrity (tension-integrity) is a term coined by Buckminster Fuller. He was exploring the balance of forces between tension and compression. He was a passionate idealist and visionary and was always interested in what connects. The famous icosahedron is a simple example of a tensegrity where Fuller had suggested "everywhere the same energy." Buckminster Fuller, an architect, inspired a way of working with energy-efficient structures. This understanding has also led to how buildings can be organized to manage earthquakes with an ability to dissipate the forces.

Kenneth Snelson (born 1927) was an installation artist who built many "floating compression" structures with a combination of alloy tubes or rods pushing and connected at points with cables. The combination of cables and tubes were constructed in a multitude of ways keeping the tubes suspended on the cables and not touching each other. This design can be created in any shape or form and is dependent upon the length of cable and tube to create a balance of forces that is light, energy efficient, and relatively flexible.

One of these installations inspired Stephen Levin, an orthopedic surgeon, in the 1970s, to understand a specific dynamic tension between bones to keep them suspended, as had to be the case with the enormous tails and necks of dinosaurs such as brontosaurus,

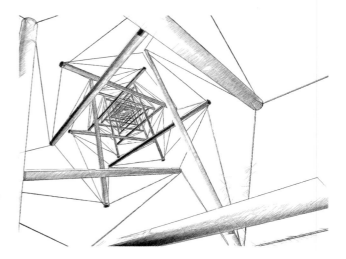

diplodocus and many others, for example. In his understanding of physical structure, it would be impossible for these creatures to hold themselves up without the dynamic of tension/suspension balanced with compression—the necks and tails of these enormous creatures would be simply too much for them to support.

Levin saw one of Snelson's installations and saw exactly how it was suspended, with struts suspended by cables. The dinosaur neck bones were suspended by soft, strong tissues under a balance of forces.

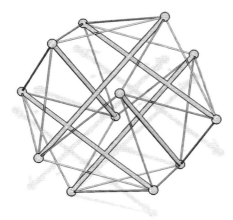

The basic tensegrity model, the icosahedron, may suggest this and as explored through various mediums, engineering and architecture, stiffer cables allow for more bounce in the system.

Tensegrity suggests resilience; the ability to absorb shock forces and disperse them with least effect on the system. If there is a delay in the structure, stiffness in a place not usually stiff, adhesion, densification, disorganized direction then the forces cannot be dissipated quickly enough and there is the potential for possible injury.

Certain tensile forces can create a catapult-type effect at specific regions around the body where forces can be managed, loaded, captured, and then released. The tensional qualities of the tissues display this potential well, loading and releasing to varied degrees. If the load is well placed, and the supporting tissues will maintain it, then energy released can be sent to where it's needed, creating an efficient motion through the structure.

"Everywhere the same energy" is also a quote by Buckminster Fuller; however, it is not exactly the case in the human or animal body. Energy is dispersed to different levels at different regions, some softer and looser, some firm, some hard, some stiff, some tight, some twisted and spiraled. The variations are enormous, yet it all comes together to create one whole action, a whole body event.

To display this a little more realistically, the icosahedron would require some elastic bands, then a guitar string, thin elastic, cling film, a balloon, stiff coat hanger wire and a spring, suspending rods that would have a twist, a curve, round ends and flat structures; together it would operate appropriately, with some structures short, some very long.

This would best demonstrate the enormous variety of the forces at play on our bodies, and the balance between them, all tensioned, as they should be to create the quality of ease and poise.

We can also think of a balloon filled with a fluid or a gel-type fluid: we can push into it and it rebounds. This is a combination of the fluid pressure and the tensioned balloon skin. Too little fluid and the balloon may still be elastic but would not rebound so well, expand it too much with fluid and then the tensile quality of the balloon can be compromised. This too is a combination of the tensional and compressional forces that are displayed in our form.

The myofascial slings and continuities

As we understand from the field of fascial research, each structure of the body is wrapped in fascial membranes, from delicate capillary structures, nerve fibers, wrappings around organs, sheaths and tendons and joints, and varied sheaths and wrappings to the muscular tissues. These are composed of collagen, elastin, and reticulin fibers all within a dense fluid base known as ground substance.

The element we are particularly focused on here is the myofascia. Myofascia is the name given to the connective tissue that surrounds, divides, and contains all the muscle tissues of the body. The fascial tissues penetrate the muscular structure through to the individual muscle fibers ensheathing each region, bundle, groups of bundles and eventually, the muscle body itself. This also includes its assorted tendinous endings and how those blend toward the skin of the bone known as periosteum. This is where the myofascial body and the osteofascial bodies meet.

It is thought that in reality we may have one muscle, folded multiple times around our skeletal structure, with moorings at key regions around our bony structure to create optimal motor, minimal compression, and balanced tensional forces.

As summed up simply here,

> *"We are a complex piece of origami."*
> —Thomas Myers

Anatomy, of course, has divided this up into individual muscles and this is a model for us to follow and supplies a useful map of positioning and basic local possible functions. Myofascial continuities that we now see present are once again an anatomizing of this structure by connecting the individual parts together in lines from head to foot and again is an excellent model for understanding tensional balance and the skeleton.

Here we create another model—and that is all this is—following the models idea, following the pathways shaped in the tissues as the anatomy trains notion follows the grain and continues at that level with no abrupt angle change or sudden depth change.

If we follow pathways of motion, shapes, and directions in the tissues when looking at the leg, for instance, we can see a vast array of directions that all work (we hope) collectively to create an easy motion to the limb. Through specific methods of fascially-oriented dissection interesting patterns, pathways and shapes arise.

The nature of the fascial tissues, the laydown of fiber directions and varied densities to the tissues, determine not only the shape of each myofascial unit (namely a muscle), but will also determine the directional forces and loads that it will transmit, respond to, or redirect.

Load capture

When considering a sling as a closed structure or a catapult, loads can be sent into the tissues and because of their structural arrangement or architecture, they can send the forces kinetically onto the next sling, or even continue the forces back to the body again.

One example of this is the classic lower region of Tom Myers's Superficial Back Line where we see the calcaneus sits in a strong sling-type structure of the Achilles tendon to plantar fascia. This can redirect an enormous load from the foot to the limb and body above—consider a sprinter on the starting blocks, or the heel of a cat or a dog. This catapult, sling shot-type arrangement captures the load and stress, passing it onward into the body, ideally, in an energy-efficient way.

Next we will explore further structures that may assist in this redirection and managing of forces.

> *"Dissecting out myofascial tissues takes one simple approach … we slide the scalpel sideways."*
> —Thomas Myers

Hamstrings as a myofascial continuity

As we know from the Anatomy Trains theory, the Superficial Back Line is a series of linked myofascial units running from head to foot or foot to head, including the plantar fascia, fascia of the calcaneus, Achilles tendon, gastrocnemius and soleus, popliteus, hamstrings, sacrotuberous ligament, sacral fascia, erector spinae group to splenius, and scalp fascia to brow ridge. However, what if we consider the hamstrings themselves as one myofascial structure? If we include the sacrotuberous ligament, the ischial tuberosity fascia onto the common hamstring tendon and follow the semimembranosus and semitendinosus it takes us to the pes anserinus.

Following the line of strain through the fascial tissues, it is then possible to once again take this direction to a common convergence around the tibia and tibial tuberosity, which is a wide (aponeurosis) around the anterior part of the tibia; if we follow this round it takes us toward the fascial sheath over the uppermost "origins" of the tibialis anterior. Following this shape and direction will make the strain pattern evident and obvious with a bent knee onto the connections of the lower biceps femoris as if its tissues blend with the fibular head and spread to the lateral fascia of the tibia. This can be removed as one complete interconnected myofascial sling (see dissection images).

This gives almost the appearance of a catapult, with a double-ended function, once again completing a series of slings suspending the femur.

This is potentially a tensile elastic structure storing potential kinetic energy. If utilized well in motion it can take the preference away from possible quadriceps and hip flexor overuse, helping to prevent knee injuries and hip injuries. Maybe not a hamstring but a "hamsling"!

been separated from the pelvic structures including the pubic symphysis and from the linea aspera at each femur; we have one entire structure; the tissues are found to be continuous from one side to the other.

These tissues are one piece and take up a large portion of the thigh. They are heavy and are organized in a gentle arrangement that allows for the ranges of motion the adductor structures can allow for. However if we consider this "butterfly" type arrangement as one whole, it becomes a strong support structure that dynamically can handle movements and positions such as various lunges and wide stride squatting movements, with a behavior like a trampoline or hammock, a place to rest a little more load to help take the extra work away from the quadriceps and hip flexing structures, which would apply more load and strain than necessary at both knee and hip.

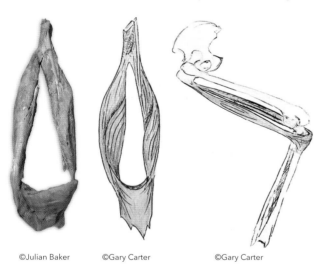

©Julian Baker   ©Gary Carter   ©Gary Carter

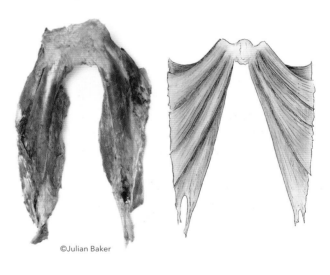

©Julian Baker

The adductor butterfly

Once again following the grain of the myofascial tissues and continuing the notion of "turning the scalpel sideways" at the ischial ramus, pubic ramus, and pubic symphysis we find the entire myofascial unit of the adductors from pectineus, adductor brevis, adductor longus along with the adductor magnus; this entire myofascial structure along with the pubic fascia has

New explorations through the structure

Here we explore ways in which some movements may load and engage with these myofascial arrangements in a few practices and disciplines from yoga, functional trainings, martial arts, sprinting, t'ai chi to the Soma Move system from Sweden. These are just a few of the movements to which we can apply a different viewpoint.

A partial squat or chair pose in yoga

Moving through this position or "gesture," an understanding of the hamstring sling can be a useful place to settle into. Allowing your weight to rest slightly (about 10%) into the myofascial tension of this sling has the potential to take excess effort away from the quadriceps, which can sometimes add unhelpful load toward the knee and hip. By deepening through the rear of the knee and sending the sitting bones back and away from the knee the practitioner is able to pre-tension the myofascial tissues of the hamstring sling.

Four point crawling

This is a position we see in all sorts of functional training systems and it is used in conditioning for sports and martial arts. Once again the pre-tensioning of this sling by allowing an amount of rest into it can prevent excess compression at the hip joint and can aid more of a free-flowing buoyant motion.

Sprinting, from blocks

Here it is possible to load the sling tissues, once again pre-tensioning, where the sling from the plantar fascia to calcaneus to Achilles tendon and triceps surae are included along with the hamstring sling. Adding just the right amount of tone here can allow for an explosive energy-efficient action through which the rest of the leg and body gather the forces to move under force and spring from the starting position.

The lunge position

Very often the hip flexors, including the poor old psoas, are focused on a little too much and can distract from the freedom, uplift, and integration that this position and gesture can offer.

Here we can pay attention to the "adductor butterfly" along with the hamstring sling. Once again, by allowing some rest into these tissues (about 10% of weight), giving the weight as if in a hammock can create quite a different sense of freedom across the underside of the pelvis and a possible feel of lift from within the pelvis, rather than the "stretch" and "pulling" sensations that often occur across the hip joint in this position.

Horse stance and warrior poses

Again, coming from a little more rest (10%) into the adductor myofascial arrangement and hamstring sling arrangement can take the emphasis away from the hip flexors and quadriceps and prevent them from doing more than they need to. This quality in the positions/gestures can assist in generating more

resilience, longevity, lightness, and efficiency toward any movement that would arise next from this gesture.

Soma Move practices

Exploring the fluidity of such practices as Soma Move and fluid martial arts forms and understanding how to engage proprioceptively through these myofascial sling arrangements can make these movements easy, fluid, free, and able to generate elastic energy in our system.

References

Müller DG, Schleip R; Fascial fitness, catapult mechanism, elastic recoil of fascial tissues. In E Dalton; *Dynamic Body: Exploring Form, Expanding Function*. Oklahoma City: Freedom From Pain Institute, 2011.

Earls J; *Born to Walk: Myofascial Efficiency and the Body in Movement*. Chichester: Lotus, 2014.

Fukunaga T, Kawakami Y, Kubo K, Kanehisa H; Muscle and tendon interaction during human movements. *Exercise and Sport Science Reviews* 2002, 30(3):106–110.

Grundy JH, Fischer, MO; *Human Structure and Shape*. Graz: Mouritz, 2015.

Hoyt DF, Taylor CR; Gait and the energetics of locomotion in horses. *Nature* 1981, 292(5820):239–240.

Huijing PA; Muscle as a collagen fibre reinforced composite: a review of force transmission in muscle and whole limb. *Journal of Biomechanics* 1999, 32(4):329–345.

Ingber DE; Tensegrity and mechanotransduction. *Journal of Bodywork and Movement Therapies* 2008, 12(3):198–200.

Kawakami Y, Muraoka T, Ito S, Kanehisa H, Fukunaga T; In vivo muscle fibre behaviour during counter-movement exercise in humans reveals a significant role for tendon elasticity. *Journal of Physiology* 2002, 540(2):635–646.

Kram R, Dawson TJ; Energetics and biomechanics of locomotion by red kangaroos (Macropus rufus). *Comparative Biochemistry and Physiology. Part B* 1998, 120(1):41–49. Doi: 10.1016/ S0305-0491(98)00022-4.

Lesondak D; *Fascia: What It Is and Why It Matters*. Edinburgh: Handspring Publishing, 2017.

McDougall C; *Born to Run: The Hidden Tribe, the Ultra-Runners and the Greatest Race the World Has Never Seen*. London: Profile Books, 2010.

Myers TW; The "anatomy trains." *Journal of Bodywork and Movement Therapies* 1997, 1(2):91–101.

Schleip R, Klingler W; Fascial strain hardening correlates with matrix hydration changes. In TW Findley, R Schleip (eds); *Fascia Research–Basic Science and Implications to Conventional and Complementary Health Care*. Munich: Elsevier Science, 2007, p. 51.

Åsa Åhman

Åsa is a certified Rolfer™, Rolf Movement practitioner™, and yoga teacher. She works with these modalities to help people improve health and wellbeing by bringing awareness to habits of movement that unconsciously create tension in the body, by breaking these habits and replacing old repetitive unconscious patterns with new more conscious ways to move that let the body hold its inner space as it moves and interacts in the space outside the body.

Is it possible for gravity to move through the body more freely? Can you move with less effort? Awareness and a more detailed body map are two key components to make changes in your posture and way of moving.

Åsa works with individual clients at her practice in Stockholm as well as with groups. She also presents workshops all over Sweden.

CHAPTER 18

What is movement?

Åsa Åhman

"Movement is the song of the body."
—Vanda Scaravelli

I see movement as the body's way of expressing itself; of getting what it needs and wants. The variety of ways we humans implement this are as many as there are humans; we are all unique. As I see it, the life that we live shows itself in the body, as if the body tells the story of the person. Not always easy to understand and see but it is all there.

I have the luxury of seeing many bodies in my work as a Rolfer and yoga teacher, both in private sessions and in groups. I continue to marvel about the strategies that we humans come up with to cope with all contents of life, both the good and the bad. We fall in love, we are happy, we experience accidents, traumas, cultural beliefs, fashion, parental advice, etc. We create habits out of what we do, just because we do not know any better. We always do the best we can, and the body always handles any given situation the best it can. How the body responds in any given moment may not be the best for long-term health, and may not have even been the way we had intended it to respond, but it always does the best to meet the requirements of the situation. Of course I also see this in my own body. I can understand my own journey, where I am right now, and why. I see some of the issues I am struggling with, habits I would like to break— but not all of them. For that I need an outside perspective from someone else. But I am now more aware and work every day on improving my awareness and perception about what is me.

My journey

As a child I was not interested in sports or physical activity. I was very content with playing—which in Sweden at that time was quite physical: biking, running, building tree houses, etc.—and reading books. My father forced me out to go cross-country skiing in the winter, and PE at school made me play soccer and dodge ball. It took until my teens to start an organized physical practice: karate. For me that was the doorway to other practices. I needed to improve my cardio to survive the karate practice. We did summer training outdoors, including lots of running. For a number of years I was the only girl training at the club. When we were training outside in the summer the group would be waiting for me to catch up, and when I did they continued. I never got a break, but my cardio improved rapidly. I also began to understand that I needed more strength and flexibility. I can't say I was much into stretching back then but at least I started working out in a gym to build more muscle. Finding the gym got me to my next step: group fitness. Aerobics! I loved it! Great music. Fantastic instructors that made me feel that I could move to music. I did every class that they had, often one class after another. Moving in rooms clad with mirrors, however, was not the best thing for me and I got quite obsessed with what I looked like, and what I weighed.

My academic background was a Master of Computer Science and Engineering, and I went from high school, selecting the top program, and then to Sweden's top university for engineering, pushing myself in the area of education as well. As soon as I got my first job (as an IT consultant) I took two weeks' vacation to educate myself to be an aerobics instructor. After that I started teaching group fitness after work as a hobby. I did that for a number of years and with my strong physique I could handle lot of exercise without injury so I pushed and pushed. I trained to become a personal trainer and somewhere around that time I tried my first yoga class.

Finding yoga

I had no idea what yoga was all about. I remember my first downward facing dog, and can still hear my inner dialogue, "What is the purpose of this? It is not improving my strength. It is not cardio. Why bother?" Luckily for me I gave it another try. Perhaps some part of me understood that this type of training was so alien to me, I had nothing of the sort in my life at all, that there must be something there that I also needed. Yoga eventually led me to a different way of moving, a different motivation and understanding. For a number of years, however, yoga was just another physical activity, something else I could practice, improve in, and perfect. I did not have a teacher that really saw me, saw what I was doing to myself and what I needed to do differently. I just did the best I could. So I pushed, and pushed some more. Yoga did not make me more flexible, but I got a better and more detailed understanding of my body and a greater body awareness. I learned to observe my mind and I eventually started calming down. Pushing less. Feeling more.

I spend many hundreds of hours training in Anusara yoga, a style of Hatha yoga based on Iyengar yoga but adapted to tantric philosophy by its founder, John Friend. This style of yoga was heavy on alignment cues, requiring a lot of knowledge of the body and the philosophy of the instructor. I no longer agree with this style of yoga but it made me a better teacher, being able to give out a lot of information—as well as asking many questions—as an instructor and teacher.

Finding Rolfing®

During one of my many yoga trainings in the USA (I did most of my Anusara trainings there) I heard about something called Rolfing. I sat behind an athletic woman at lunch one day and heard her say that she was often complimented for being so strong but that there was a price to pay for all that strength. She said that she had had to do the Rolfing 10-series twice in order to start getting more flexible. "Hmmm, I thought. I am quite tight myself, maybe that is something I should try."

Back in Stockholm I found a Rolfer and booked my first session. At the first session the Rolfer worked on my breath and helped me feel my breath in my back body. At one point I thought he had slipped me some kind of drug because what I felt in my body was something completely new. He worked on the right side of my torso, front, back, and sides, and had me stand up and walk after that. It felt like my ribs protruded about 10 cm more to the back on the right side compared to the left. It felt so strange and new to me! Then he worked on the left side and I felt more balanced. He sent me home with the homework of continuing to explore the breath in my back. I booked a next session and then completed the 10-series. I felt immediately that this was something I wanted to learn more about but at that time I did not have the time, or money, to send myself through the Rolfing training. I just pushed it to the back of my mind and continued with my life. But the thoughts of Rolfing would not go away. I got to a point in life where I felt really content. I was working part-time as an IT consultant and part-time with exercise and health. I liked having the experience of these different types of work. The IT work provided one type of universe with its people and aspirations, the gym, with personal training and group fitness classes, another. Yoga added another dimension completely. I was excited and happy to be involved with all three diverse ways of working.

Listening to my body

Even though I liked where I was in life, my body started protesting. I did not realize it at the time, but all the movement I was doing just reinforced my existing patterns. I used the same muscles over and over again to do any type of movement, not the muscles I thought I was using. I released more and more in the same places as always to meet new requirements for flexibility, rather than getting more flexible all over as I thought I was doing. I started getting pain in my body and for the first time I started to be more selective in my exercise. I stopped doing certain classes and moves because I knew they were not good for me, but I was clueless about what to do instead, to support and strengthen my body in other ways.

I know now that I was always making too much effort in everything that I did, not only with exercise. I did too much of the same thing. I fed my patterns of tension. In the way I was practicing it, yoga did not make me more flexible, rather I just got better at controlling my body; it made me stronger. Pain forced me to stop doing certain things (running, for example), then stopped me from

teaching certain classes at the gym. I realized it was not healthy for me. For some time I was not sure what to do instead. Having no way of breaking my patterns (my awareness was not that great) I did not want to risk making my patterns stronger. I even thought about stopping yoga, since it also just fed my patterns, but instead I started shifting my practice and teaching into something that suited me better. The thoughts of Rolfing persisted. I realized that my patterns of always pushing a bit too much also was keeping me in a profession that was not for me. Just because I could do something well was not reason enough to stay doing it. So, during this time I finally quit working with IT and enrolled in the Rolfing training and followed my newly acquired knowledge about movement and pattern breaking. I am still doing that. The Rolfing training and now also working with clients, has helped me understand more, and I am still learning.

What about your journey? Are you aware of your own patterns of movement? Or in life in general?

Components of movement

One way to look at movement as a tool to break habits is to look at its two components:

1. Giving weight to the ground.
2. Adding a direction. A wish. A need. An intention.

For any kind of movement, we need a starting point or reference point. Something that stands still so something else can move. Somewhere to put the weight, to push off from. It seems so simple, and we all do it, but it can be quite powerful when you actually put some attention to this.

Imagine taking a single step forward with your right foot. Try doing that without first dropping your weight into the left foot and leg. You will not be able to take a step! You might shuffle, or jump, but not step. Giving weight to the ground is something we do all the time, but where we choose to give our weight is not always the most optimal, and sometimes we even resist letting go of our weight. Talking about weight can be quite abstract though, and it is not something most of us are used to doing. Sometimes it can help to think of us humans as water-filled creatures. More or less 70% of our body is water. Now try to imagine pouring your weight, as if it were water, into that left leg.

When your weight has dropped into the leg and foot, and you add a direction, from a wish, a need, or an intention, then you move. You give your weight to the ground, and then want something, or need something, or want to express something, and there is movement!

In order to change habits and the way you move it is worth the effort of paying attention to HOW you give your weight to the ground—it is not only about doing it. Can you simply let the mass of your body be affected by gravity? Not completely, of course, for then you would fall as a puddle on the floor. Notice where it is possible. Can you feel the weight of your pelvis? Chest? Head? Arms? There is quite a lot of weight to an arm but it is something we are so used to carrying around all the time that we are desensitized. Since we do not feel their weight, it is easy to start lifting the arms with unnecessary effort from the wrong place, quite commonly seen as, but not limited to, the upper trapezius. Other strategies can also be made up and found.

One thing that can support dropping the weight into the ground is a specific kind of sensitivity called *hapticity*. The term comes from dealing with newborns and how they touch and perceive but has been adapted by Hubert Godard[1] to describe a dual form of awareness in touch. One of my Rolfing teachers described hapticity as "the ability to touch and be touched at the same time." When your feet stand on the ground, not only are the feet pressing into the ground, the ground also presses into the feet. When

we get curious about what we touch, the feet meeting the ground, for example, the body has less resistance to fully connecting to the ground. So if you can, take off your shoes and socks, stand up straight on your two feet, and imagine that your feet are trying to "see" what they are standing on, as if they could even tell the color of the floor or ground that you are standing on.

Giving your weight to the ground, to have something stay with the ground, is a prerequisite to have something else move more freely. This action is called *phoric function*. Phoric comes from the Greek word *phoro*, which means, "to carry." It is another term coming from Hubert Godard, and it is something that we already do, more or less consciously. But the more we are aware of this phenomenon, the more we can move delicately, more in tune with gravity, and with less effort.

Rolfing and yoga: different methods, same goal

"Strength that has effort in it is not what you need; you need the strength that is the result of ease."
—Ida Rolf

Rolfing and yoga need not have anything in common, depending on the style of yoga that you compare it to, but for me, they both have the same goal in mind. This is to have a body that is:

+ free of unnecessary tension, with freer movement patterns and habits
+ adaptable to its surroundings and relationships
+ more capable of expressing itself, and getting what it needs and wants.

Describing what yoga is would take up a couple of books by itself, but the types of yoga that we see in the west normally include *asanas* (poses and shapes created with the body); *pranayama* (breathing exercises); and *bhandas* (techniques to control the energy flow in the body), for example. Other things can be included as well since the development of yoga is deeply intertwined with Indian culture and religion.

I see an opportunity to use the rich treasure of the yoga asanas, together with awareness and breath, to practice inhabiting the body and mind in such a way that we understand ourselves better in terms of habits, patterns of thoughts and movement to evoke positive change. This is the same thing I want for my Rolfing clients.

So, what is Rolfing?

Rolfing® structural integration is a holistic and process-oriented method for health and wellbeing. The method was created by Dr. Ida Rolf in the 1930s. She was interested in how to use the elasticity of the fascia in regards to medical care of physical problems. She regarded gravity as a healing force. Interestingly, Ida herself started manipulating bodies with yoga before she got more "hands on." She apparently called it *yog.*[2]

Rolfing works with the fascia of the body through manipulation (pressure and massage techniques), movement, and education to allow gravity to flow through the body in such a way that it nourishes the body instead of being something we fight.

Injury, stress, and trauma affect the fascia, making it shorter, thicker, and less elastic. Movement patterns such as long hours in front of a computer screen affect the fascia in the same way.

Rolfing helps the fascia regain its elasticity, vigor, and vitality so the different parts of the body can move better independently but also better together, using less energy to move, to sit, to walk, and to breathe. Because you are more efficient in your body use, you have more energy to use on other things.

Rolfing aims to go to the root of your symptoms, which may present as postural challenges or appear as pain. Everything you have in your body, including tension and pain, is there for a reason. Rolfing helps to reclaim the knowledge or free movement that your body has always had, but which has been hidden and forgotten behind layers of habit, a posture your culture or context tells you to have, one-sided movement, stress, etc.

What to do?

To help the body organize itself better in gravity is to help it move more efficiently. Even without going through the Rolfing process, we can all use the principles behind Rolfing to improve our movement, break our habits, and get a freer posture with more possibilities. Together with the postures, breath, and awareness from yoga we have powerful tools to help release tension and realize another way of being in our bodies.

Three questions

One simple thing to observe and do is use the idea of the components of movement that I described earlier, namely weight and direction, but, to make it more practical, to see it as weight and lift. Ask yourself these three questions:

1. Where can I feel weight?
Where can you feel the pull of gravity? Where can you feel the mass, or watery content, of your body? When you can more clearly feel your weight you are better prepared for question number two:

2. How do I lift?
Unless you are lying as a puddle on the floor you are lifting yourself up somehow. But where? And how? What

do you do to stay up? And when you find your lifting action, can you do it more easefully, with less effort?

Try this for an exercise: stand upright. (If you know your yoga, do *tadasana* or mountain pose.) Stand in a neutral stance with head high and a long spine. Then, let yourself feel gravity and relax into it. Go as far into relaxing as you can without falling to the floor. In order to do this you need to forget everything you know about good posture! Let yourself slump. Let the shoulders be rounded. Allow the knees to bend or buckle. Let the head hang. Find the minimum amount of effort to still keep you on your feet but not much more. Stay there for a breath or two and then, slowly, start lifting yourself up again, but not using your normal strategy. Do it slower. Maybe you can imagine yourself being lifted up by an outside force? Can you find a more gentle way up where you use less effort than normal?

We all have our patterns of lifting, habits so well used that we often do not notice that they are there, especially when it comes to lifting. That brings us to:

3. Where am I holding?

Can you, with exercises like the one above, start to recognize places where you always "lift," or actually hold, because you do it constantly? I call them *scout muscles:* muscles that are always ready, always prepared! They jump in and help to lift and stabilize, even when it is not their job. My well-trained scout muscles are in my shoulders, especially the right one. If I do not pay attention I can exercise my left arm and my right shoulder wants to "help." If I need to use my abs, I promise you, my shoulders will go into action, but since I am now aware of this, I can change this pattern—and have done so. Not always, but more often than not, I can see this pattern activate and choose differently. The first step to making such a change is first, to become aware that it exists. Then, for some (or at least for me), a long and really boring internal dialogue follows, "Oh, I feel my shoulders are tensing. Can I relax? Yes I can. Great." Ten seconds later, "Ok. Tension again. Can I let go? Yes." Ten seconds later I am there again. But with some patience the ten seconds become 20, then 30, then a minute, and then longer and longer. Now they only kick in when I do something really strenuous or new, or if I get really nervous. (Ok. They still kick in quite often!)

Mary Bond describes this beautifully in her book *The New Rules of Posture.*[3] We over-stabilize for various reasons such as pain, fear, or just by lacking orientation (more on this later). These patterns, when successful, then become habits; habits that we keep even though the original danger is long past. Our adaptable bodies become less adaptable, shorter, compressed, with less space in the joints, fewer possibilities to move. Our dynamic way of responding to gravity and life diminishes and we fall victim to gravity. We start to feel the pull down without being able to have a buoyant response in the opposite direction of up.

Orientation

Whether on the yoga mat or off, another way to work to help the body hold less tension and have more possibilities is to help it orient. The body can take care of most things, move beautifully without our mind interfering (and by that I mean our ideas on how movement should look or feel) if it knows the answer to one seemingly simple question: where am I? By that question I do not mean geographically, I mean where am I in regards to the ground and to the space around me. Kevin Frank has written an excellent article about tonic function[4] (another term that comes from Hubert Godard). The muscles that support us, that lift us up, need to be oriented in regards to ground and space to function properly. For example, your hamstrings, which are supposed to be tight because they are part of those muscles that hold you up, will lengthen as much as you need if they have clear direction of up and down, unless of course the tissue is not functioning properly. But, if the tissue itself is pliable enough, all necessary length is there if the body knows its directions.

Using the senses

As children we are less concerned with control of our movements and just go with our needs and wants. We are guided more or less by our senses—what we see, smell, hear, or taste. Touch is another strong sense but I will save that one for later.

If we can let our senses be wide and open then we are in a good place for free movement. When the body is under stress, in fight or flight mode as we call it, when the sympathetic part of the autonomous nervous system is activated, we focus on one task only: survival. All the senses narrow in order to focus on the enemy to fight or to run away from. In contrast, when we are relaxed we are open to the world around us. If we feel safe our peripheral view is active, sounds and smells are noticed from both near and far. Are your senses wide?

Eyes

Where you are sitting or standing, relax your shoulders and let your eyes look straight ahead. Keep the skin around your eyes soft and imagine that your eyes have deep roots that go all the way to the back of your head, as if you were looking from deep inside your head and not from the surface of your eyes. Then, without moving your eyes, be curious about how much you notice of what is on your left and right side. How wide out to the sides can you see objects or the room you are in? You can even use your hands and arms as a way to measure your sphere of sight. Start lifting your arms out to the sides, slowly. Gently move your fingers so they are easier to see. How far back and how far out to the sides can your hands be with you still seeing the movements of your fingers? Keep moving your arms and hands in a big circle over your head to check the view all the way up over your head. Don't be surprised if your view is not symmetrical. It is quite common for peripheral vision to differ from left to right.

Not only is the peripheral vision active when you are relaxed, engaging the peripheral vision can be a way of becoming more relaxed! Try this when walking to help you stay grounded. Can you see the world moving past you, to the left and right, as you are walking forward?

Ears

Using my hearing as a way of keeping myself oriented is quite a new tool for me but I have found it to be an effective tool. It can be difficult if you live in a big city where there tends to be so much noise close by that it drowns out everything that is further away but if you happen to be in a less crowded area, try this for orientation:

Listen for sounds that are close by, but also for sounds far away. Go deeper and deeper with your hearing. What is the sound that is furthest away that you can hear? Maybe a church bell in the distance? Or a car driving by? Wind rustling the leaves of a tree?

Nose

As with hearing, the sense of smell is challenged by living in a city. Then, of course, the human sense of smell is not particularly well developed, but it is still useful to help you orient. Let your nose be open for scents that are close by but more importantly for scents that are coming from far away.

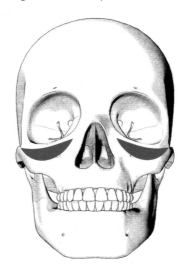

One way to make the sense of smell stronger is to connect the nose to your vestibular organs. With a strong and clear touch, use the base of your hand, from the thumb to the heel of your hand, to mark the pathway from the nose through the cheekbones to your ear. Hold steady for a minute or two, pressing in and with a slight lift. Then go out into the world pretending that you are smelling your way onward, as if you were a dog with a broad and well-functioning nose.

Using hapticity

A feeling of weight is one way to be connected and oriented down, and what really helps with orienting both down AND up is hapticity. If I can be curious about my contact with everything that is around me, everything that is not me, then the body has a better understanding of where it is. We are always in relation to someone or something. We are always in relation to the outside. If not, there would be only me and that universe would be rather boring. We need, however, to be in relation to ourselves as well as to the outside. Sometimes in yoga, or sports, or life, we can get too focused on the outside. All that exists is what is outside of us. It can be our boss at work, assignments at school, the image in the mirror. If all our focus is on the outside then we lose ourselves and that is a shortcut to a burnout. In yoga class it can manifest as trying to copy a form as perfectly as possible, or following a teacher's instructions about alignment, or looking into the mirror and trying to get the lines and angles perfect.

Sometimes, we can go the other way, and be completely self-absorbed—all that exists is me and my feelings, my needs. If that happens, then you are not that fun to be around and not really functioning in this world. We all need the ability to stay in relation with ourselves at the same time as we are in relation with the outside. I think that if we practiced this more the world would be a much better place and many of the health issues we are dealing with now with modern illnesses such as burnouts and different stress symptoms could be avoided.

In the beginning you might have to keep oscillating between the outside and the inside. You check in with you: how are your stress levels? How is your posture while you sit at work? And then you go outside again and focus on the task ahead. With practice you can become better at being both in and out simultaneously. Can you do what is required of you and take care of you, or at least check in with you, at the same time?

Use hapticity to also stay connected with the space around you. Feel the air touching your skin, or feel your clothes. Can you be aware of the space above your head? Or behind your back?

As Mary Bond describes in *The New Rules of Posture*, when the body is not free to orient we compensate by stabilizing too much. And then we fall back into the bad loop of holding too much and shrinking our bodies and closing them. So to keep the body open and expressive, we need to be oriented.

An example: downward facing dog

So, let's look at a couple of different ways to approach a very common pose in physical yoga, downward facing dog. It can look and feel very different depending on how you do it and what your focus is. Let's try it using two of the ways of helping the body that I have described so far:

Dog pose with weight and direction

Start on your hands and knees, hands underneath your shoulders, knees under your hips, toes tucked. Take your time to feel the contact with the hands and metatarsals on the ground. If you could not see what your hands and feet are touching, could you guess only from the contact? Is it warm or cold? Soft or rugged? Is it moving or still? Take your time with this investigation to help the body relax into the ground and for the weight of your body to drop more into the support from below. Can you feel the weight of your pelvis? If you shift your pelvis over your heels, what happens to the weight in your hands? What happens to the weight in your feet? Before we add direction through the tailbone, make sure the two ends of the spine are free; check that you are not holding any tension around your neck and jaw. Also check that you are not clenching the

back of your pelvic floor and blocking your tailbone to find its direction. Now, feel the heaviness of your head in one end of the spine and imagine that someone pulls your tailbone back and up so your body lifts up into downward facing dog. Can the neck stay soft and head heavy during the lift and when you get up? When you have reached up high with your hips and tail, can you imagine that your arms are hanging from your back? Can you also imagine that your legs are hanging from your hips? So that they might get even heavier, not because you are pressing hands and feet down, but because you drop their weight more clearly down into the ground?

Before you release the pose just check how your breath feels while being in dog pose like this.

Dog pose with clear orientation

Start on your hands and knees and take some time to breathe. Follow the breath and what it does to you as it flows through you. What happens to you when you inhale? What happens to you when you exhale? Or rather, what happens to you when the breath comes to you and when the breath leaves you? Breathing is not an activity; it is an experience, a movement. Can you perceive it as such?

Try closing your eyes and be really aware of what is beneath your palms and feet. What does the surface feel like? Is it soft or hard? Warm or cold? Smooth or rough? Does this contact change with the breath? How is the contact on an inhalation compared to the next exhalation? Try to keep this curiosity going. And what about the room and space around you? How does your contact with that change as you breathe? If this is difficult to imagine, picture the air as somewhat thicker, almost as if you were under water, or that the air wrapped around you like a warm blanket. How does your contact, your relationship to the space around you change with the breath? I call this to have a breathing relationship to ground and space. If you pay attention to the breath you also have a breathing relationship to yourself, since the breath also flows within you.

Then open your eyes again and let your senses guide you as well. Use a soft gaze with active peripheral vision and listen for sounds that are close to you and sounds that are further away. Build your sense of the room around you, mapping it with as many of your senses as you can.

To come into the pose, stay in this wide awareness and curiosity, both within and without you, on what is you and what is not you. Check that your head and neck are free to move and that nothing is holding around your tailbone. Let the pelvis be heavy over your heels and imagine the tailbone being lifted back and up. As that happens feel how the contact to the ground changes. Imagine that your feet and hands are sinking deeper into the ground as the pelvis gets lifted, and observe the head hanging heavy and free at the other end of the spine. As the pelvis gets lifted, the sit bones (with their hamstring attachments) are moved away from the heels. So with one direction going down through the sinking feet and one direction going up through the lifted pelvis, the hamstrings stretch by themselves.

Stay in dog with wide focus on the orientation down, through haptic hands and feet—a contact that changes with every part of every breath, and orientation up and out through haptic skin to the space around. That contact also changes all the time. Like any relationship.

Not this simple

Of course nothing is ever as simple and well categorized as we would wish. To best help the body orient we use a bit of everything, always. Rolfers tend to talk about a triangle of perception and orientation: feet, vestibular system, and eyes. Of these three the eyes are quite often the strongest. Many of us try to orient by grabbing the world through our eyes. To soften the body, or to de-stress, let there be more balance between these three, at least. Let your eyes be a guide, but include soft haptic feet. Maybe check that the head and tail are free to move?

See what works for you. Perhaps you can use my text as a way for you to unfold your way of orientation. I am sure you will find that some things work better than others and different tools are suitable for different situations. Play with it. Explore. Make it yours. Make it useful to you.

What's next?

I think that playing with the three questions (How is my weight on the ground? Where and how do I lift? Where do I hold?) can help you break and renew any patterns of movement. Practicing yoga, or anything in life with orientation—using the senses, hapticity, etc.—could maybe be the best relationship practice in the world. When does a human relationship take a turn for the worse? Most often that happens when we stop being curious about the other and we stop being curious about what happens with us when we meet that other. So, if I can practice staying curious about the ground beneath me or the space around me at the same time as I am with me, then I can handle any relationship!

When I look at my own life in the rear-view mirror I see how I have shifted from doing and pushing into more sensing and being open. Ida Rolf is supposed to have said, "Over and over again, people come to me, and they tell me 'you just don't know how strong I am'. They say 'strength' and I want to hear 'balance.'" I used to be all about the strength. Now I am more into balance: helping myself and others into a more balanced body and way of moving.

I am curious as to how my own journey will unfold in the years to come as I, hopefully, will continue to learn and sharpen my sensitivity to myself and others even more. This text is a reflection on my understanding of movement right now. I know I could not have come to this understanding without going through the steps of my life just as I have done. I also understand that this level of understanding is not a stopping point, but just the next step on the way. The journey continues as long as I stay curious. Let's stay curious!

Note and references

1. Godard H; PhD. Professor of movement and research, University of Paris.
2. Rolf I; *Rolfing and Physical Reality*. Rochester, VT: Healing Arts Press; 1978, p. 7.
3. Bond M; *The New Rules of Posture: How to Sit, Stand and Move in the Modern World*. Rochester, VT: Healing Arts Press, 2007, p. 8.
4. Frank K; *Tonic Function, A Gravity Response Model for Rolfing Structural and Movement Integration*. Rolf Lines, 1995.

Julian Baker

Julian Baker has been a Bowen therapist for 25 years and a teacher of Bowen since 1994. He is the author of two books on Bowen, *The Bowen Technique* and *Bowen Unravelled: A Journey into the Fascial Understanding of the Bowen Technique*. He is one of the world's leading experts on Bowen. After studying with Gil Hedley, he began to study fascial anatomy to better explain Bowen and, now leads fascial dissection courses at medical schools throughout the UK.

His writing and presentations attempt to present complex subjects in a simple language that is easy to understand and follow; as a result, he is regularly asked to speak and lecture around the world. His Functional Fascia company (functionalfascia.com) teaches fascial theory and dissections.

CHAPTER 19

Fasciaism is on the rise!

Julian Baker

 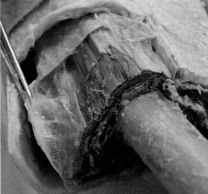 

These are some examples of the wide range of shapes, forms, and textures that the fascia present in the body.

The last ten years have seen a flowering, rise, birthing, of the fascia movement, with conferences, papers, treatments and Facebook pages dedicated to all things fascial. From an anatomical perspective this has been an interesting development. It is widely accepted within anatomical circles that fascia is generally poorly understood and described, has for the most part been cut away and discarded or ignored, and is considered by many the poor relation of the connective tissue (CT) family. Its more recent rise to fame, however, has been accompanied by some dubious claims and a lot of misunderstanding, and there remains a whole part of it that is still overwhelmingly neglected.

All connective tissues have three things in common. Firstly, they all arise from the mesenchyme, a type of connective tissue that develops from the embryonic mesoderm. The cells of the mesenchyme give rise to all the connective tissues, blood vessels, blood cells, the lymphatic system, and the heart.[1]

Secondly, connective tissue is going to have a wide variety of blood supply. Skin for instance has lots of blood supply, whereas cartilage has comparatively little.

The third element that defines connective tissues is that they are all mostly inert, or non-cellular. It is important to remember here that non-cellular or inert does not mean dead or non-living and it is also a good idea to point out that although they are mostly inert, there is still plenty of cellular activity going on.

The main element that is most important when it comes to considering the role of connective tissue in movement is that all connective tissues, although low in cellular content, are abundant in intercellular substances and fluid. It is these intercellular substances that contain an abundance of fibers and which require movement and sometimes loading, in whatever form, to distribute and organize them.[2,3]

The classifications of connective tissue give us two types of connective tissues in the body: (1) specialized connective tissue, which includes blood, bone, cartilage; (2) connective tissue proper, which has the subcategories of loose connective tissue and dense connective tissue.

Blood can sometimes be omitted from the category of connective tissue, as it lacks the fibers that some authors deem is a requirement for a true connective tissue, but it seems to be an important aspect of connectivity so we're going to keep it in our connective tissue basket.

It is also worth noting before we go much further that there is quite a variation in opinion when discussing connective tissues in general and an even bigger divide when the subject turns to fascia.[4] There is, however,

quite a lot of information out there about most of the CT types, so for the purposes of this chapter, we are going to concentrate on connective tissue proper.

Naming parts of the body has always meant that there will be bits that will get less attention or even get left out completely. In the case of fascia and some other connective tissues, the classifications could be considered incomplete at best and there are plenty of other tissues that have been ignored completely.

This gives rise to the idea that if they haven't been named in the last 500 years of anatomical categorization, they can't be that important now in the 21st century—and good luck trying to prove otherwise.

However, there is quite a lot of classification that exists already and in order to get to grips with what we are talking about, it is worth taking a brief tour through some of the tissues to find out what different ones do, where they are, and how they fit together.

Fascia as a stand-alone word does not feature strongly in any of the categories, but instead could find a home in any one of the categories we are looking at.

Connective tissue proper

Loose connective tissue (LCT)

The main feature of loose connective tissue, compared to dense, is that it is less fibrous, has more cells in it and contains more ground substance, a component of the extracellular matrix.

ECM and ground substance

Extracellular matrix (ECM) is the most abundant and diverse tissue in the body. It makes cell movement, cell communication, and all normal function possible. It protects the cells that surround and produce it and is flexible and incredibly adaptable. Yet at the same time, it is an inert and non-living tissue, produced by cells (extracellular is the clue).

Ground substance is the starchy, mucus-like liquid that is present in the ECM. Ground substance contains a lot of fluid and it is through this fluid that cells and blood vessels can obtain nutrients and dispose of waste.

It consists of starch and protein molecules, mixed with water and held in place by fern-like projections called proteoglycans, which are proteins with molecules of carbohydrate added to them.

These starchy strands form a flexible mesh that can trap water and combined are referred to as glycosaminoglycans (GAGS), or mucopolysaccharides, "muco" suggesting mucus-like and polysaccharide meaning lots of sugars. Hyaluronic acid is one group of GAGS and is a major component of synovial tissues and fluid. The main job of GAGS is to stabilize the structure of collagen fibers in the ECM.

It is this mesh-like fluid structure that allows for compression, force transmission, cell communication, fiber development, electrical and energetic impulse transmission. You name the function and you can be sure that the body probably could not do it without the ECM and ground substances at their core.

There are three main types of loose connective tissue:

1. areolar
2. adipose
3. reticular

Areolar is the most common connective tissue and means "small open space." It is irregularly arranged and consists of collagen and elastic fibers, a protein polysaccharide ground substance, and connective tissue cells such as fibroblasts. Its function is to hold organs in place and allows epithelial tissues to attach to other tissues.

It also acts as a feeding and waste station, containing as it does macrophages and white blood cells; nearly all cells get their nutrients from, and dump waste products into, areolar connective tissue. It has lots of characteristics, is full of ground substance, is found around nerves, muscles and blood vessels, and forms the connection of the skin to the underlying fascia.

Adipose connective tissue consist mainly of cells rather than ground substance, but it is held in place by more dense connective tissues, which we will come to later. Adipose, in spite of the hatred that our society holds

for it, is vital for survival. It is both a storage house for energy and an insulator as well as providing a padding and cushioning for organs and the skeleton.

As well as these very practical elements, adipose might also have some surprises in store as more recently it has been proposed as an endocrine organ. Adipose makes, stores and responds to hormones, particularly those relating to appetite, with the hormone leptin being produced here. Leptin is an appetite suppressant and has a role to play in blood sugar regulation. As leptin is a hormone, this seems to suggest that adipose tissue could be an endocrine organ, something which doesn't seem far-fetched when one considers the evidence.[5]

When adipose is held in place with a connective tissue network, the tendency is to refer to this as superficial fascia, a term which we will return to presently.[6]

Reticular connective tissue is similar to areolar, but has lots of reticular fibers which are a type of collagen produced by specialized fibroblasts called reticular cells. Reticular fibers also contain, in some places, elastin. It provides a framework for blood and organs such as the liver and kidneys and also provides structure for the spleen, the lymphatics, and bone marrow.

Reticular cells are pretty widely spread out throughout the body and there are over 20 known types of reticular fiber.

A word here on the protein elastin and elastic fibers. The flexibility we achieve in the skin, blood vessels, and lungs is in large part down to elastin and many of the mechanical properties of connective tissues are determined by the relationship between elastin and collagen.

It is a beautiful relationship. Elastin provides the stretch whilst collagen, the most common protein in the body, gives strength and resilience. The strong, multiple directional collagen fibers give a foundation for mobility that the elastin can take advantage of. Think of the expansion of blood vessels and arterial walls that takes place on a constant basis.

Aging is the enemy of elastin and diseased arterial walls will also see a breakdown in elastin. In COPD (chronic obstructive pulmonary disease), it is the loss of elastic tissue in the lungs, often due to inflammation, that creates the lack of elastic recoil and difficulty in breathing that is associated with the disease.

Dense connective tissue (DCT)

As with loose connective tissue, there are also three types of dense connective tissue:

1. regular
2. irregular
3. elastic

Regular dense connective tissue is where we start to get into the familiar and recognizable territory of fascia. Regular dense connective tissue is collagen based, with reticular and elastin fibers less present. The regularity is recognizable but in spite of its classical explanation and definition it does not run in clear parallel lines. Instead, it weaves and dives in and out of its own structures, to create strength and stability as well as a ceiling for underlying muscle.

Regular dense CTs can be sheet-like, such as the iliotibial band, or cord-like, as with the Achilles. There is a lot of interest in these tissues and as a result, a lot of information out there. The question as to whether we can change these tissues with manual applications is one that creates a lot of heated discussion that still has a lot of time to run. My personal view is probably not, but the reasons for that are something we will have to address at another time.

Irregular dense connective tissue is less arranged than regular and spreads out like ice on a frosty pond. Mostly collagenous, it also contains fibroblasts and some ground substance and is located in areas where there are variable directions of tension. Examples are in the dermis of the skin, around the periosteum of the bone, and in the tunica albuginea, the fibrous envelope of the penis.

Elastic dense connective tissue is found where you need more elasticity than rigidity or force transmission. The vertebrae need to be able to twist, bend, and be mobile and it is here in the vertebral connections where we find a lot of this kind of tissue.

It is also present in the arterial walls to allow stretching and flexibility, but as can be easily demonstrated within this area, it can easily stiffen and is hugely variable and changeable according to disease and environment.

All of these types of connective tissue are present when we use the word fascia. Where you happen to be in the body and what you are asking the body to do define the type of connective tissue, and by default fascia, that you are talking about.

Bone is one connective tissue that does not come under the category of "connective tissue proper" but which has some similar properties in terms of collagen and cellular activity. It has been suggested in some quarters that bone is fascia. Starched fascia admittedly, but fascia nonetheless. It is an argument that doesn't really hold up, due the fact that bone has another job (amongst many others) of regulating the calcium levels in the body, which places it in a category of its own. If bone were starched fascia, then starched fascia could do the job of bone. It can't.

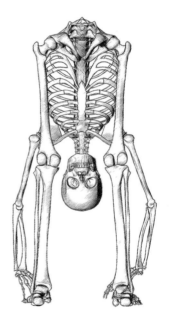

Connective tissues are connecting tissues

It is a statement that sounds pretty obvious, but in the process of studying, classifying and analyzing connective tissues, it is often easy to forget what they actually do.

When we talk about layers of the body, it can be misleading. The layers we see in the process of dissection are those that are created by a very sharp knife being used on a dead person. They are a sculpture, a creation of something that does not exist until we bring it into existence.

Each structure we discuss or talk about can often be seen as something that is present or existent as a disposable part or element that has no relationship to the other things we've talked about. A car has lots of parts. You can buy bits for it, take them off, throw them away, buy different sorts of parts and items for it.

A body is not like that, and it is this approach and mentality that has created such a huge problem in medicine, where specialists in various bits of the body and even the mind abound, but where no one looks at the whole unit and, to the best of my knowledge, there are no movement consultants employed in hospitals.

This does seem to me to be a missing link when you consider the statement that there is no disease mechanism where there are not changes in the ECM. Unless and until formal health systems embrace and incorporate movement into every layer of health care delivery, we will be unlikely to see much in the way of progress.

The superficial fascia

Now that we have briefly examined all the tissues that make up the connective tissue system, we are going to drill down into one particular area that we can all see, feel, and touch, in an effort to understand how this one, very neglected and poorly discussed tissue contributes to our understanding of movement.

The tissue in question is the superficial fascia. You might notice that this term does not get included in the above classifications and descriptions of connective tissue. That is because there is no real consensus as to what constitutes superficial fascia or where it should be categorized.

Often referred to as subcutaneous, it is mostly referred to as either superficial adipose tissue or deep adipose tissue, the two being one and the same structure. If at this point you are confused, then you are not alone. Trying to extricate oneself from the myriad of terms and explanations is something of a nightmare task.

So for the moment let's keep it simple but also understand that the way I will describe and talk about superficial fascia is not an accepted scientific or anatomical approach, simply because there isn't one, or at least not one that is universally accepted.

We have already seen that adipose connective tissue mainly consists of cells; adipocytes to be specific. Adipose or fat cells are laid down as energy storage around the body and in the last few years we have seen a lot more laid down than in the last 100 years.

There are lots of potential reasons for this, but essentially the human body has virtually limitless storage capacity in terms of calories. We can keep eating and taking on calories on the assumption that at some stage the calories will stop and we will enter starvation mode, where our energy stores are going to be needed.

This is all very well when our food sources are based on seasonal access to mammoths, but becomes a problem when the local fast food joint is open 24 hours a day seven days a week and is happy to deliver. The big culprit here seems to be sugar, which has a sneaky way of bypassing our natural appetite control systems, but in any event an excess of intake will end up being represented in the adipose.

Adipose, or for that matter any kind of cellular structure, doesn't just hang around by itself, but needs some kind of container in which to exist. In the case of muscle tissue, it is the fascia that creates the container within which it can function. The adipose tissue also needs a connective tissue container and so is held in place by a network of collagen fibers and in turn is inextricably held in place by the boundaries of the skin.

The superficial fascia network is body-wide and, irrespective of the amount of adipose that is held within it, is present in every human form. There are of course places in the body where adipose is less prevalent and places where there is a lot. But any time you pick your skin up, with a pinch or any kind of movement, you are also, by default, picking up superficial fascia.

Skin does not move around on top of deep fascia, but is always attached to the superficial fascia, with the exception of the eyelids and certain parts of the genitals.

On the back of your hands the layer is thin and around your abdomen or upper leg it is going to be thicker. This is a pattern repeated around the body. A couple of mms in some areas and several cms in others. The thickness and location of it will depend on several factors.

Genetics will play a large part in where you lay down your adipose tissue and there is absolutely nothing you can do about this. There is an idea somehow that measures of beauty, or even health, are determined by the measurements you have around certain parts of your body. We are told that certain exercises or movements will help to burn fat in these areas.

Nothing could be further from the truth. If you want to know where your fat will be laid down, look at your parents. If you think that certain exercises or diets will "burn belly fat," ask yourself this question: when you were eating that biscuit, did you direct your digestive system and metabolic process to store that biscuit in your butt? If there are exercises that can burn your belly fat, why aren't there exercises to burn off the fat in your wrists? Or knees?

The idea is of course ridiculous but it buys in to a lot of people's insecurities. Superficial fascia is the presence of adipose tissue, embedded into collagen fibers which hold these cells in place. It is impossible to get rid of manually and it can only be eradicated by metabolic processes that involve hormonal responses. These are predominantly based around a change in the amount of food and energy we take in, coupled with the amount of energy we expend.

It is not always that simple, as there are genetic and environmental factors that play a role in fat absorption that are just being discovered. Our complex and emotional relationship with fat still has a lot to play out.

Liposuction that tries to suck the fat away from certain areas will also need to take the collagenous tissue that is part of it, and will involve a heating process and a cutting or slicing action in addition to suction to remove the offending tissue. The result is that it could just come back, but without the mesh that holds it in place, the effects could be unpleasant and the space left could—and sometimes does—just fill up with scar tissue and less organized fat.

The idea in people's head is that the fat in their thighs or belly is somehow like a packet of butter that can be melted down and sucked out. Most people have no understanding of the fat being held in place by a mesh which needs to be cut away along with the fat.

The superficial layer can be thought of as a biological wetsuit. The tissue itself is a poor conductor of heat, creating an insulating effect that works both ways, keeping heat both in and out of our system. Thermoregulation is vital for our system to function, and the first call on this is the skin and superficial fascia layer.

In terms of movement, we can use the wetsuit analogy again. Try putting on a wetsuit that is a size too small for you and what will happen. Whilst it won't suffocate you and you'll still be able to move, you will feel severely restricted and held in place.

The skin and superficial fascia need to be able to move and glide over the surface of the deep fascia in order for normal functional movement to take place. If it gets stuck or pinned down for any reason, such as surgery, injury, scarring, or inflammation, then restriction is going to be experienced.

The trouble is that this restriction is often attributed to other issues, mainly because the layer of skin and superficial fascia, particularly as a moving functional tissue, is poorly described and misunderstood and all too often dismissed as fat.

A simple way of demonstrating both the existence of these tissues, as well as experiencing their role in functional movement, is to pinch some tissue from the elbow at the end of the biceps. This is an area that has plenty of slack skin for the precise purpose of allowing a highly mobile joint to be mobile.

Hold your arm at 45 degrees with your palm face up and pinch some tissue from the crook of your arm at the front of the elbow. Hold on to this as tightly as possible and what you have is a wedge of skin and superficial fascia in your fingers. Now try to extend your forearm away, whilst at the same time keeping hold of the pinch of skin. In order for this extension to take place, the skin and superficial fascia is going to have to be pulled away from your grip.

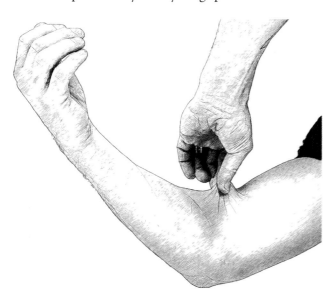

If the grip is too tight, then the arm won't move and in the process of trying to make it move, a lot of pain in this soft tissue will be experienced.

Around joints and mobile structures, this seems fairly obvious and easy to demonstrate. As long as we keep moving our joints around then we shouldn't see this kind of limitation. But what about in areas where we have less in the way of "natural" or functional movement?

I say functional movement because a lot of the movement we discuss is often "extra-functional." Functional is the movement that is part of a natural, everyday requirement, whatever that requirement might be. Anything else is extra-functional.[7]

Someone reaching up to get a cup from the cupboard is performing within their functional range. Asking them to lean backward in a yoga pose would be, for them

at least, extra-functional and would need practice and application to achieve.

The average 55-year-old person might therefore not be expected to have a huge variety of movement of the superficial tissues in certain areas where there is less joint mobility or movement, such as around the abdomen, lower back, back of the neck and so forth. Similarly the amount of adipose deposited in these areas is likely to have an effect on how much movement there might be.

It is hard for most people to imagine the idea of movement in this layer being relevant and even harder to describe so I was incredibly excited when this year I saw the most practical demonstration of this I had ever come across.

The process of dissecting skin as one distinct layer, followed by the total removal of superficial fascia/adipose as another, was something I have done many times and is an idea pioneered by Gil Hedley and continued in my dissection classes. In certain subjects, the superficial fascia can be removed in one complete fleece-like unit, whereas in many if not most, it is much harder.

In this class we were able, with the dedication and skill of the students in the class, to take off a generous layer and lay it side by side with a layer of superficial fascia removed from a very thin donor. Both were female.

This image demonstrates the extreme differences that can exist in the superficial tissues of the human form.

The larger donor, however, had received a double mastectomy and there was significant scarring on her upper chest wall across her skin and around into her shoulder and pectoral region. It would not be unreasonable to assume that this level of scarring would create a pinning effect onto underlying tissues and this is a common presentation around areas where a surgical intervention has taken place.

Whilst the skin does not happily let go of its connection to the superficial fascia, resisting its removal at every stage, the superficial tissues are much more mobile and fragile. For much of the area it is possible to use a blunt dissection technique (fingers) to clear away the loose connective tissues attaching superficial fascia to deep.

In this case, however, the superficial fascia was completely adhered to the deep layer of connective tissue and had to be physically cut away at every stage during the removal process. This adhesion was not just present in the area around where the surgery had taken place, but virtually everywhere throughout the upper body.

Even after removal of the superficial tissues, pulling on the area of the scar also created a movement through the head and neck, suggesting that whilst alive, this movement may have created a tensional force through the chest area.

It is a dangerous thing to make assumptions about function and movement, based on dissecting people who are no longer functional or moving. We did not see this lady when she was alive and have no idea about how she moved or did not move—for all we know she could have been a gymnast to the day she died.

What we can see from this is that even in cadavers we can mimic the restriction that occurs by physically pinning down the tissues and attempting to make various movements through a limb or other area.

The beauty of the dissections I oversee is that we use a specialized preservation technique that allows for a full range of movement on our cadavers. At the beginning of the process, we examine and palpate our donors and do this at each stage throughout the dissection process. When we have removed the skin we always see a wider range of available movement and once we have removed the superficial fascia we again see an even greater range.

We can therefore reasonably hypothesize that the arrangement of skin and superficial fascia over the deeper layers of fascia and muscle is a considerable contributor to the range of movement that is available in normal function.

Disturbance to this arrangement, whether it be through surgical intervention, disease, immobility, or lack of normal function, will create a change in the distribution and cellular arrangement of the extracellular matrix and the ground substances. In other words, stiffness.

This understanding has long been a central tenet of structural bodywork and movement therapy. Something is stiff so we move it and it becomes less stiff. The assumption until now has been that this stiffness occurs at a deeper level in the body, around the joints or in the muscles, associated with lactic acid build-up and so forth.

The concept of more superficial tissues being a contributor to movement or lack of it is perfectly logical if the comparison of the wetsuit or external restrictor is applied, yet has not gained traction in the bodywork community.

The reasons for this are relatively simple to understand. Muscle, bone, and to a degree skin are all areas of study that are commonplace within the field of anatomy and therapeutic intervention. The superficial layer, full of adipose tissue, is still just considered to be a fatty structure, with little or no relevance to movement and merely something to be reduced by dieting and exercise. It is rarely considered when discussing movement or studying human function and even in the academic world of anatomy and physiology, it is generally cut away and ignored in order to get to the deeper tissues.

Medical students and students of anatomy will rarely see a larger person as a donor in a dissecting room, these donors having been turned away at the point of death due to their size and generations of practitioners never seeing what really lies beneath their hands.

The practical applications are relatively straightforward once the theory and relevance of working with this tissue is fully accepted. It needs to move. If it can't move by itself it needs to be moved by someone else. Picking it up along with the skin and moving it gently over its underlying connections changes the way it responds. There is an increase in blood that comes to the surface of the skin and there is a change in the fluid dynamic of the tissue.

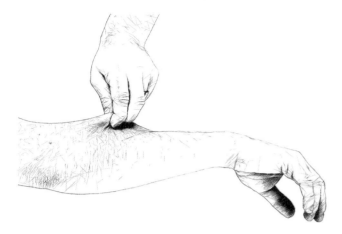

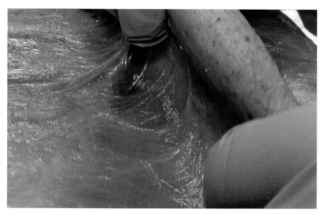

For many, there is a change in pain reporting response and an immediate increase in ranges of movement, even where no diagnosis has suggested any form of superficial restriction. Movement against gentle traction of the skin and superficial fascia is another aspect of this form of superficial mobilization and is simple to perform and practical for people to do at home. Stretching alone is unlikely to have much influence on the superficial tissues, unless some surface resistance is applied. It is my view that much of the response from taping techniques comes from a change in the tension of the superficial tissues.

The spaces that run between the collagen and adipose interface are referred to as interstitium and have been recently credited with a wide range of functions and importance,[8] even with the suggestion that these channels could be a pathway for the metastasis of cancer.

It is not hard to see the structures and tissues that make up our surface as being integral to our sense of self. When we see another, we do not recognize them from their bones or their muscle composition, but from the features that define them. Their skin and their fat layer are the primary identifiers. Movement is an inside to outside, constant, endless wave-like action and there is no part of us that is not inextricably involved with every movement we make and every breath we take.

References

1. Van den Berg F; The Physiology of Fascia, in R Schleip et al. (eds); *Fascia, The Tensional Network of the Human Body*. Edinburgh: Churchill Livingstone, Elsevier, 2012, pp. 149, 151.
2. Ibid.
3. Schleip R et al. (eds); *Fascia, The Tensional Network of the Human Body*. Edinburgh: Churchill Livingstone, Elsevier, 2012, pp. 149, 151.
4. Wendell-Smith CP; Fascia: an illustrative problem in international terminology. *Surgical and Radiologic Anatomy* 1997, 19:273–277.
5. Kershaw E, Flier J; Adipose tissue as an endocrine organ. *Journal of Clinical Endocrinology and Metabolism* 2004, 89(6):2548–2556.
6. Lancerotto L, Stecco C, Macchi V, Porzionato A, Stecco A, De Caro R; Layers of the abdominal wall: anatomical investigation of subcutaneous tissue and superficial fascia. *Surgical and Radiologic Anatomy* 2011, 33(10):835–842. doi: 10.1007/s00276-010-0772-8.
7. Lederman E; *Therapeutic Stretching*. London: Churchill Livingstone, 2014, p. 287.
8. Benias P, Wells RG, Sackey-Aboagye B, et al.; Structure and distribution of an unrecognised interstitium in human tissues. *Nature: Scientific Reports* 2018, 8(4947). [Online] https://www.nature.com/articles/s41598-018-23062-6 (accessed May 23, 2019).

Cecilia Gustafsson

Cecilia is a yogi, dancer, and entrepreneur. She has worked in the movement business for many years and is a well renowned teacher and presenter. She has developed several movement concepts, of which SOMA MOVE is one. Cecilia works with a holistic perspective in everything she does and is always searching to deepen her understanding and "connect the dots."

CHAPTER 20

What is movement?

Cecilia Gustafsson

Movement is everywhere. An ongoing motion in universe: seasons, days and nights, life and death.

The human body itself is a never-ending movement, with different rhythms from the tiniest cell to brainwaves and the slower lymphatic system. The opposite, stagnation—loss of movement—is always experienced as unhealthy, wherever it occurs. In economy, bodies, feelings, thoughts, and relationships interconnect. In other words—let's keep things moving!

In my movement work over the last two decades I have come to find four qualities that I try to encourage people to explore and embody, in order to have a happy movement life. These are grace, rhythm, power, and stillness.

Grace

"Learn how to think with your whole body."

Grace—this lovely quality of bliss, beauty, and dignity has a touch of art and elegance over it!

Even if you often can see grace in movement styles such as dance, gymnastics, or martial arts, you do not need to be a dancer, gymnast, or martial arts practitioner to seek and perform it. Grace is a quality that can be applied to almost every movement in life. It is rewarding and desirable and worth looking for.

When moving in grace, you move with ease, elegance, and sensibility in a receptive and present way. Your techniques become silent and look effortless, it feels like you are listening and thinking with your whole body. And for someone who is watching, your movements look harmonious and beautiful.

It is about presence, timing, and using the right amount of energy for the task. So, not trying too hard or being insensitive like a machine, but instead being perceptive about your situation, the surrounding world, and your inside world: then putting it all together in a "wholeful" way, and keeping in touch with your underlying intention.

Balance

"Not a thought. Just an intense presence in the placement of my body, that with small adjustments rediscovers the balance over and over again. A small tuck in of my tailbone, a lift in my neck, a slightly greater press into my fingertips, an opening in my chest, the awareness of my back and that I could lean a bit more toward it. And then my left foot, pressing down to the floor and meanwhile my right side rises higher. I'm in a yoga position called Vasisthasana. A floating feeling. Weightless. A process of trying and failing has led me here. Every time an adjustment in toward the balance point. Adjustments that become smaller and smaller, more subtle."
—C

Balance is not something that you find once and then hold. It's a permanent adjustment to not losing it that requires commitment. Balance is mirroring our actual presence, steadiness, and alertness, as well as our contact and interplay with the different parts in our body.

If we are too tense or not in touch with ourselves, we usually wiggle if we try a "yoga tree" position. But if we then make ourselves do a few balance positions we can reconnect to a centered self. While its almost impossible to think at the same time as we do a balance challenge, we get an anatomically "here and now" presence. You become where your body is now, not in the past, in the future, or in another space.

We all have a front, a back, an inside, and an outside. An upside, downside, a right and a left side. Be aware of that when you move. Don't lose contact with the less obvious side or your roots if you want to rise … A balancing act is like a dance between these polarities.

When you try to find your balance point, in life or in movement, you first explore and touch the edges, maybe tipping too much first to one side, then to the other, before eventually moving closer to the center. It is the same for sailing, surfing, riding a horse, making a pirouette, doing a headstand—when you get perfect lines, things get easy, with less resistance. Your body and your movements will be steady and easy at the same time, like a golden tunnel. Balance is good.

Posture

"Root to rise."

Imagine a tree, with branches that reach toward the sun, a trunk that stands steady and strong above ground, and under the same ground a root system that is as large as the crown. A beautiful visualization of posture between the polarities to expand and to be grounded.

I like the quote "find your roots and wings." A reminder to honor your depth and respect who you are, and where you come from, while you move toward your vision, longing, and freedom. Try to let those words and the tree image move into your body next time you stand, walk, or are doing your yoga, dance, or training practice. This will give you a secure grounding that holds you steady and centered, and it will give your posture an extension—a positive feeling of freedom and being uplifted.

Root down your feet and legs like the tree, let them find their way down into the ground while they draw nourishment and energy into the rest of the trunk, your core. Feel how the energy is transported through the perineum and up, how it lifts your spine, solar plexus, and heart, making your back strong like a trunk and your front open. Like the crown toward the sky and the light, your arms want to expand and explore and your head is light and carried proudly.

Remember: if you only want to fly you fall easily, and if you forget your wings you get too heavy. It is simply about a dance between roots and wings!

Presence

"Arms and legs are moving across the floor in a rhythmic flow, now and then I spin around my own shoulder and feel how the gravity catches me, sucks the fluids toward to the skin. The blood is bubbling inside me, my feet are light, sometimes so light that I almost lift off. My breathing is pulsing perfectly in interplay with the heart and my mental observation and awareness fills me without disturbing my body. I move in the room as if making love to the air around me, and my eyes are open but are not seeing anything outside my own zone. Everything flows, movements release each other, and it seems as though I am discovering new movements I have never done before. I rise and fall in intensity, everything from emotional or physical impulses. My brain is remarkably still, no thoughts of upcoming moves or analysis, just a mental awareness that observes and supports. Sometimes I direct the light of the presence toward a detail—the sound of my feet when they meet the wooden floor, the feeling in the chest or the sight of my hands' own dance through the air. Most obvious is the feeling that I want to stay here forever, in this rich, colorful, restful, and timeless space. A space—a state of sweetness and flow, where I am not limited by standards, thoughts or time."
—C

The treasure is hidden within. When you open yourself to the presence of now, you will be rewarded.

Flow, the Zone, or Satori are different names for the golden state where you surrender yourself to an activity or mediation, release excessive control, intellect, anxiety, self-consciousness and become more transparent and present. There are no thoughts of future or past, no resistance or ignorance toward the now. It feels as though someone or something is holding the steering wheel and you will be released from your ego, yourself, and thoughts of your life for a moment. It has an intoxicating feeling of concentration, timelessness, and euphoria. An inner kind of sweet reward instead of an outer.

The ways to get to this presence are several, but those through the body are the most available.

Unfortunately it seems like many of us have forgotten that the body is a source of pleasure and deeply satisfying sensual conditions, and instead directed ourselves toward outer goals like improving health, accomplishments and a good-looking exterior. That's a pity, because if you learn to improve the quality of your bodily experiences you can sow seeds for those juicy, transparent moments that enrich your life and you intuitively are longing for. Here are some keys for receiving flow through your movements:

1. The activity you choose to do has to be challenging or interesting enough that it requires your commitment and focus, but not so difficult that you become stressed, too afraid or exhausted. You should have confidence that you can manage it if you rise to your higher self.
2. You need a mental focus that is well balanced—like a rider on a horse. A rider that supports and leads, but without inhibiting the horse's own movements.
3. Use a limited focus glance. In that way you can increase your "inner seeing" and your other senses.
4. And last, don't look for the flow—plant the seeds that could bring you there and lose yourself in the moment and movement.

Sensibility

Don't forget to condition the ability of sensibility, to stay alert, awake, and receptive. This quality is blunted by too much tension, as it is by stress, anxiety, and anger, so the first thing to do is to soften a bit so that the natural sensitivity will increase in the feet, hands, and mind. Then you become a sentient being rather than a robot on autopilot, and you will move with much more elegance.

The hands and feet in particular are equipped with a lot of neurons—they are carriers of information, they talk to you, about texture, temperature, and ground. Listen to them. Use them as receiver and sender toward the floor and the world. Let the hands be open and alive, so you can feel if you need to put a little more pressure forward or back, to the right or left in order to keep balance. In a handstand, for example, you are adjusting your whole body from your fingertips and wrist. With too tense or dead hands, that would be impossible—just as it would be impossible to tell the weight of a stone, or the fine texture of a silk dress or a lover's cheek. In order for the feet to carry information to the rest of the body, let them be barefoot in your practice if possible. A bare foot is much more sensitive and fun to do movements with. With shoes on, we become clumsier and not as well balanced, and we will also lose contact with earth and the experience of grounding. Besides, it is very beautiful with active bare feet where the toes feel, stretch, and work like they were a being of their own.

To be sensitive is also to be able to perceive signals, not only the obvious ones but also those that are more quiet and small—within your body and senses, as well as in relation with other people. Be able to perceive and respond to nuances of emotions, subtle gestures, and delicate experiences, to have gut feeling. That is unfortunately what can get lost if we just "go for it" with grit and great force. Make it a habit to control if you contract (close, tighten) or expand (open up, radiate) when you do your practice, meet people, are thinking of something special, etc. Most often it is best for us and a more positive experience to choose the expanding state.

Rhythm

"Energy moves in waves. Waves move in patterns. Patterns move in rhythms. A human being is just that. Energy, waves, patterns, rhythms. Nothing more, nothing less. A dance."
—Gabrielle Roth

Rhythm is our mother language. A language we can trust and which takes us back to our creative nature and ecstatic humanity, with roots all the way back to the absolute beginning. A language that goes beyond intellect and sense that goes straight to soul, flesh and spirit—the spark we inherit through all the generations.

We are born with a natural will to move in rhythm. As small children we rock, swing, and express what we hear with our bodies. We are born knowing the sounds of the heartbeats of our mother, first in the womb, and then when we lie on her chest—therefore the sound of beats has a physical anchoring with us and is experienced more by the body than with the hearing.

Musicality

Rhythm is pleasant. It is a balance of emptiness and content that can move us forward to more power, inward toward a nice trance, or down into relaxation. To be in rhythm makes everything, especially movements, less exhausting. Much of what you do can be done with an intention to be made in rhythm—not forced, not inhibited. You may be writing, walking in a crowded street, flirting, driving a car, making love, having a conversation or running, dancing or doing your vinyasa yoga. Even breathing.

Try to open yourself to musicality next time you do something with movement.

Hear or feel rhythms everywhere, the waves, the patterns, and the pulse. Feet against the road, hands to the floor, inhalations, exhalations, heartbeats, arm movements through the air, a movement. Let the body parts communicate with each other like musicians, where the cooperation becomes a musical piece. What should be played? Rock 'n' roll, symphony, or soul? Sometimes when I move through my day, I have a soundtrack following me everywhere. I become a part of it, move, think, and talk in rhythm with that real music—try it!

Back to nature

"Breaths that rise and fall, expand and retract. My head is heavy, falls to one side and the other with the hair sweeping over face, neck, and floor. Bare feet are stomping, hips shivering, and the heart is pounding in rhythm. Movements as round and circular as the blood. Intensity changes with impulses from within. I can discern my hands' warm heaviness when they pass my face and I feel intimately close to that which is rhythm, gravity, and bliss—life and origin."

The sound of a drum, a heavy beat, reconnects us to nature. To the more primitive parts of ourselves. Where the rhythm lives, where the hip is moving spontaneously. Everything that connects us to nature seems to be healing for us. Those who have tried to surrender to music in a dance know that something happens there, which makes us more grounded, more human, more connected. The soul is rolling up and out from hidden nooks and seeps out in skin, arms, and feet. We can scent that dance is a universal language that is waiting to be expressed. When life has been too digital, artificial, and contracted—then one needs to reconnect to flesh, sweat, breath, and beats. Throw off your shoes and socks, release your ponytail (if you have one), and put on some music you can feel in the chest. Let arms, spine, feet, and head start to move in circles and rollovers. Seek the form of rhythm that is round and sliding, not straight and clean-cut, like marching. Dance until you sweat, then dance a bit more. Let the feet find a way, maybe tentative at first, then more confident. Let the hips swing, the pelvis tip forward and back, slow or fast. The spine rotates and winds like a snake, the neck is following and your arms are leading you upward, downward, backward, or sideways. You are dancing. No rules, just a language that starts to express itself, and the fastest way to revive body and soul. After half an hour you are somewhere different to where you were before, and you never knew beforehand where it was going to be.

Would you like to dance even more?

Try this dance exercise at home if you want to explore what different rhythms and movements could give you. You don't have to be a dancer in order to dance.

Just like you don't have to be a runner or a gardener in order to run or plant flowers in your backyard. You will find that your resources of steps and movements will get bigger and more advanced every time you explore the dance.

Do it like this. Read and learn the following guidelines and let them be a support for your practice. Happy trip!

1. Start

Awaken your body parts, one by one. Put together your feet, arms, head, and spine in a dancing cooperation. Let your awareness shift like a strobe toward each body part. Start with your hands, move them as if it was the first time that you felt them, or saw them. Move on to elbows, shoulders. In how many ways can you move them around? Let yourself become aware of the joint, how it gets lubricated, opens. Let your spine wake up. Get to know all the ways it can be moved. Get into every little corner of it. The back becomes the hips. Sink down into them as if they were a container for the upper body. Let your feet and legs get involved. Open and contract your ankles, step around, feel your legs taking you around and across on the floor. Still your feet, and let the head roll heavily side-to-side with compassion for your neck. When you feel done, you are done.

2. Fluid

Choose music that takes you to a swinging rhythm, resonant with a bit of soul. Explore the heaviness and gravity in your movements. Anchor yourself down low in your body, breathe and increase the sense of being in the now instead of thinking of the next move. This is round lines, flow, with no starts or stops. Get yourself into patterns that feel comfortable. Stay there for a while. Stay on one hip. Explore how you can angle, circle, and drive the hips, like you were rotating you tailbone down to the earth. Can you make a spiral go upward from your thighs to your neck? Like a water vortex from a bottle. You can shift the tempo whenever you want from fast to slow or the opposite. Enjoy the earthy, motherly rhythm like fluid in your blood.

3. Sharpness

Get a song that gives you beats, sharpness, and cockiness. Explore angles and straight lines. Starts and stops in the movements. Forth, back, side-to-side. Imagine you are doing karate, shadow patterns. Exhale now and then. Feel how the movements are awake, alert—like you on a great workday. Imagine you are painting with your right hand through the air. Making lines, crosses. What color?

This dance is not for spectators. It is made for your own experience. You are allowed to look ridiculous. That has no meaning here. Chop, cut, punch. Cut whatever you don't want to carry anymore. Frustration, cute smiles, your ex-husband's choking energy. Make patterns on repeat. In that you will find meditation.

Now let your left hand become a brush. Paint with that left hand in the air. Expressive lines, like you were scattering red color on a white wall. The eyes are sharp, your body as well. Cut whatever holds you back.

Still yourself and observe your breathing and pulse.

4. Passion

The music could be a drumming, trance-dance kind of song, a bit wild and primitive with a rhythm that has a drive forward. Your theme here is to let go, let go, let go!

Let go of control, tension, and old bitterness. Shake the body like an animal does when it leaves an unpleasant situation. Some of us are good at leading our body—do this, do that. Now it is time to let the body lead you. It decides what it wants to do, you follow. Relax the jaw, the neck. If you want to close your eyes, try that for a while. Close your eyes and open your heart. This is a heart dance. It shall beat, the blood shall circulate! Move closer to a high pulse. Don't think. You don't have to. Nothing is to be analyzed. If you paint with your hand now, what color is it that is splattering around? Move like a water vortex, a waterfall, with natural force and flow, all letting go. Let your arms swing like a rag-doll. Shake the body where it has to relax and heal. Become a receiver. Let go of resistance.

5. Grace

Here we can use a classical theme on music. Violin, cello, or piano. Airy. Follow the energy that says that everything is easy, bright, and well. In the body there is lightness and hands are like butterflies through air. A lifted face with soft lips and eyes. Explore spins and balances. Where is your center of gravity? Seek extension in your body. Grace. This is like dancing a beautiful text—poetry.

6. Stillness

Calm music or none at all. Let the activity become quieter. Listen more closely to the body's need and desire to stretch or open somewhere. Rock it like a baby, wag yourself into stillness. Do slow motion choreography, become completely still now and then. Listen and think with your body, not your brain. At last, become like the stone that falls through the river down to the bottom. Dangle downward and eventually touch the ground. Lie on the floor. No movements except the one inside you. The flowing blood, the risen spirit and the becalmed mind. The rhythmic within.

Rhythmic movements in stillness

Next time you do yoga, stretches or a lying relaxation, try to invite a small rhythmic movement, a soft rocking forward and back or side to side, even circular movements, small and soft, like you rock a baby to quiet it. It brings a sweet organic feeling to what you do.

Explore this even in bed when you want to fall asleep and see how relaxing it is. Maybe we respond to this because of all the fluid we contain—the water inside the body will get waves and create ripple effects throughout the body. Imagine a bottle of water that you push or circulate in a rhythmic way, how the water starts to move accordingly to that movement.

You can also connect yourself with the rhythmic dance of your breath. Like a balloon, the body will expand a bit during inhalation, there will be a lengthening action along your spine, and like the same balloon the body will get rid of tension and become more released during your exhalation. A rhythmic massage on the inside, that gently opens tension around the chest, diaphragm, stomach, and back. Can you imagine yourself being that balloon right now—being filled up and expanded, lengthened in your spine, and released and softened while you follow your exhalation. Stay with this wave for a few minutes and enjoy the vital feeling.

If you place this kind of breathing into a movement, you will find it to be like wind in a sail. Helping the movement forward, creating power and harmony. Do your inhalation while you open or prepare, and exhale to give your movement some of that wind, that energy. In yoga, the technique and science of breathing is called pranayama, and is described as the wind that moves prana—the life force within us.

In a still meditation you will also find a rhythm even though you are completely still on the outside. There is always this little movement of rise and fall of your breath—in your nostrils, inside your chest and even in other places of the body. Give yourself time to get to know this lovely feeling of being intimate, connected to your breath by following it instead of being too active about it. Just stay close, tie your awareness around it, and rock gently up and down with it like a boat tied to a buoy. When you do, you will feel less lonely—there is life inside, a friend, something you can stay close to for a while.

Cycles, a form of rhythm

Everything in nature, including human life, is constantly moving and has its own rhythm and cycles. Evolution, full moon, new moon, cocoon, butterfly, birth, death, flowers that bud, flourish, and wilt. Day and night, spring, summer, fall and winter. Can you find also cycles in your own life? In age, puberty, menopause, career, relationships, and family life. In activity, recovery, coming forth, retirement, growth, fallowness. How would it feel if you danced with these rhythms a little bit more? Neither cling nor rush things, but try to be responsive and sensitive to these natural cycles. To let go of things, people, and phases when it's time, and step into new when you are being called. Reduce a bit of ego control, fear, and resistance and continue nature's community. Trust life and maybe you will be more healthy, happy, and free from injuries.

Power

"I can feel how the fervor is growing inside me. From the lower part of my belly to hips, thighs and up toward the solar plexus. If I would describe it with color it's orange-red, if I would describe it with a feeling, it's powerful, almost sexual—not in a romantic way, but primitive and potent. My inner body is awake, the blood has a pressing flow and my hands are warm. My back is confident and I have an urge to express something physical, release this energy through my arms and legs, prove my vigor, let the current from my inner center be pushed out in movement."

To be physically powerful is fun. To experience strength, a pulsing blood and capacity. Capacity to make an impression, expression. To make things happen. Manage. To taste a fiery energy, the one that gives vitality, and a flirtation with the masculine extrovert. It is normal to become more sharp, resolute, and energetic after a workout with a bit of "heat." It resolves stagnant energy, burns up excess, and gives new energy.

Sometimes I provoke some of my yoga students to become a little angry. When they look too pale and numb, like there is too much of an introverted and yin-like energy, and too little vitality. I want the blood pressure to rise and the blood to flow in order to give them some color in their cheeks and a glow in their eyes! I push or poke them a little, like come on—wake up! It works. They become more energized and awake, happy to get out of a dull state.

Power positions: a strong body language

Within the quality of power a strong body language is central. What signals your posture and body will send, not only to your surrounding world, but also to your inner self.

And yes, you can fake it until you make it. With broader legs and arms, relaxed shoulders, an open chest and a lifted chin and gaze—you can raise your self-confidence and signal security, clarity, and presence toward others and into yourself. Just as reducing, slouching, body postures will do the opposite.

Maybe we already know this instinctively, but we can also lift it to a more aware practice due to research that shows how power positions seem to lower cortisol levels and raise levels of testosterone—which in turn gives higher presence (when we don't have to scan for threats and be reactive), more courage, and a greater chance to solve problems in a new way. (Read more about the studies by Amy Cuddy, social psychologist, professor and researcher at Harvard University.)

So, making movements with a strong body language liberates us from small and weak postures if we normally tend to limit ourselves with that. Many of my students, in Soma Move or other classes, tell me about the special feeling of liberation and primitive power that they experience from the practice, which they love! Why not use this knowledge when you stand before a challenge where you want to increase your resources of power; before a competition, presentation, or something you are afraid of.

Stand tall, broad, and reach your arms over your head—make yourself big and let your body tell you (and others) that you are a winner.

Directed, collected, from depths

Very often, we think one thing, feel another, and do a third—no wonder we feel scattered and confused. In order to give impact to what we do, we should be more congruent. Direct focus and collect the energy of our thoughts, heart, and body into the same direction.

We may want to run, make a golf swing, hit a punch, dance, have a conversation, ride a horse. Or reach whatever goal we have in life.

When it comes to movements, try to peel off what is not essential, and focus your resources of breath, energy, and intention to one strong and centered beam of power. In one of my classes I have exercises that are similar to martial art techniques—imaginary swords, punches, and kicks, where you want to go from point A to B in the straightest, fastest, and strongest way. Even the gaze is straight and you move like it was for real—if you hesitate you are lost. This is a practice in moving forward, to become distinct and powerful. When you have been moving like this for half an hour, you start to feel like you can take decisions—chop, chop—and put words into action. Most of the time there is a majority of women in my classes, and they seem to feel well when getting clarity and authority into their body language.

When you want to make a movement powerful, remember to use your deeper muscles. A strong and fast arm is coming from an action from your whole trunk, chest, and shoulder. Probably even from your hips, like a strong and fast leg has its origin from the gluteus, hips, and back. In Asian disciplines they talk about the center of the core—Hara—like a pool of energy from which you get your force, like a water hose, transported out to arms or legs. Try to use that image next time you want strength for some arm or leg exercise.

Relaxed power

Many of us make the mistake of tensing up in our attempts to be strong. One nerves oneself, clenches the teeth, and wipes out all forms of softness. But being powerful is about using the strength in the right way. You cannot be too tense, because then you inhibit the movements and consume unnecessary energy, and if you are too soft and "sleepy" there will never be any explosiveness in the movement. The trick is to stay relaxed to the extent that you can, and then you will get access to rhythm, vitality, flow, and presence, which are important resources for vigor. That place, with the right tone and alertness, is something you want to look for, both in your training, but also in your life. It is not always easy, but rewarding if you do. It is called Satori in Japanese, a word for a state that includes both power and peace.

Play with the balance to be as relaxed as you can without losing strength, check that you are not tensing up parts of your body that you don't need to, and try to find rest where you can in what you do, even if it's hard, inconvenient, or strenuous. Imagine a martial arts practitioner, dancer, javelin thrower, sprinter, or the bear Baloo—if you were in their bodies, you would feel the underlying stream of relaxed power, ready to switch on in a contraction or extension just at the right moment—not too early, not too late, but with perfect timing. As Muhammad Ali said, "Float like a butterfly, sting like a bee." Or as we say in our Soma Move practice—move like a panther!

There is a conflict in trying too hard. That will create a tension when the opposite is obvious. Try to have a relaxed attitude toward what you are about to do—not too afraid of either success or failure. To have a smile on your lips—but not tense. Look for the enthusiastic lust to master something, with the acceptance that you might fail, instead of feeling that you "have to" perform and are not allowed to make mistakes. There is a huge difference in experience, and probably better results, with less effort. Playfulness and openness can take you far.

For me, relaxation is one of my first priorities when it comes to mastering something and reaching out with what I want in different contexts, since I know that both my ideas and I become more powerful then. My body language becomes better, words come out right in great sentences and I am present enough to catch the circumstances and people around me. So I try to shake out excess tension from my body like a dog does, take some breaths where the exhalation is like a deep sigh, and ground myself through my legs and feet.

Inwardly, I visualize the desirable action and result with a clear calmness.

To use your power well, try the following:

1. Fake it until you make it! Use strong body language and power positions to give positive signals to yourself and to others.

2. You will find the power in movements from your center.
3. Focus your resources of breath, gaze, movement, and intention in one collected direction. Peel off distractions and scattered attention.
4. Look for the confident enthusiasm to master something, rather than the fear of failure.
5. When you move, imagine yourself as a panther. Alert, supple, relaxed, ready to be explosive and fast.
6. Dare to stand up in your full force, but still not self-assertive. Like a sun—warm, strong, and pleasant.

Stillness

"Yin holds Yang."

Stillness. Is that relevant in a text about movement? I heard a well-known dancer and choreographer once say, "What is dance?" He continued, "From stillness, movement develops. To prove its necessity." So in order to move in different ways, we need that still point to begin from.

In this fast, active and info-overloaded world we live in, we easily miss out time for stillness, silence, and reflection. But with no firewood there is no fire, with no yin there is no yang. If you want to be powerful, flourishing, and growing, you need that time and space to withdraw and recover. To get to know the silence, and rest in a simple being state without interacting with other people, or things, or problems. To notice there is a life beneath the acting and thinking. You won't disappear.

And you don't waste your time because the rest is full of life. It digests and repairs, opens up and cultivates. We don't just rest *from* something, but also *in* something.

When we let ourselves be passive and receptive, things can come through to us. When we haven't filled our consciousness 100% with thoughts, predetermined content, and action, new things can get in. Insights, creative solutions, a new look at something. Or things get a chance to leave. Tensions, worries, thoughts. The ordinary attention can stand back for a while and allow an extended world.

Quiet signals

Stillness and silence are required in order to perceive more subtle, quiet signals, both from our own body and emotions and from other people. We can perceive the sensations of our breath, the warmth of the skin, the happy bubbles in our chest, or the sad tension in our throat. It is required in order to reflect, and reflections are necessary if we want to develop a sense of who we are deep inside. It is a way to sit down and wait for your soul, as the Indian quote says. The soul gets a chance to be heard and join with your body, woven together in a root system that gives depth and stability.

Clear your mind

"Where attention goes, chi flows."

It is a great idea to empty your mind regularly of things that don't mean anything, to prepare for things that do.

Minds get messy. Like beds they need to be made every day.

To center yourself—to meditate—is a way to tame our jumping, scattered mind that very often provides us with thoughts that are unimportant, negative, or unfriendly. We will quiet parts of the brain that scan for threats and danger, and instead strengthen the parts that experience gratefulness, spaciousness, and connection.

To meditate is both pleasurable and functional. To find rest in a chosen focus or decide what to give attention and energy to. To be aware that you can observe your thoughts and feelings, that they are separate from you; you don't have to follow your thoughts, resist or comment on them. There is a room in the middle, where it is calmer and more quiet, in the middle of the storm, in the middle of the wheel. Go there sometimes, in your everyday life and in movements. To a still point.

Conclusion

Movement is everywhere. Outside as well as inside ourselves. It is a dance in the smallest cell to the whole universe. Motion is a way to change your emotions, lift your spirit, and raise energy. It is necessary!

Explore and enjoy—and let your body become your playmate and friend.

Lucas Henriksson

Lucas is a certified somatic educator in the tradition of Thomas Hanna. He is also a facilitator and program and curriculum developer for Tergar International, conducting meditation courses online and onsite throughout Europe. He also holds a BA in Buddhist Philosophy and Himalayan Languages.

CHAPTER 21

What is movement?

Lucas Henriksson

The quest for a definition of movement is an interesting one, and one I am not certain we have a definitive answer to. Certainly we can grab a dictionary or type "movement" on our keyboards to immediately find an answer to this question; however, I would argue this, at best, gives us a very limited understanding of what movement really is or can be.

Is movement only the physical act of moving our bodies? I think this is what most of us understand when we hear or think of the word movement—a person doing yoga, someone dancing, climbing, running, and so on. But is movement limited to the physical act of moving alone? I find this a question worth asking.

For example, if a person enters my clinic, lies down on the practice table, and I ask them to take a few deep breaths, relax, and turn their awareness inward, simply connecting with their bodies—and then, as the person continues to relax, I begin to gently move their shoulder joint—is this movement? The person is passive—in other words, there is not an active, intentional act of movement taking place from the person's side. However, from the outside, the person's body is moving in space, i.e. movement is taking place.

This then becomes a question of whether movement is an active, intentional act of moving alone, or if movement also includes passive, unintentional actions, for example, a bodyworker moving a client? I guess it all comes down to how we want to define it.

The above is simply an example to illustrate that movement can either be active or passive, intentional or unintentional. At this point, you might ask yourself, "Why on earth is this important?" I would argue it is important, because by beginning to think in this way and asking questions, we can broaden and deepen our understanding of not only what movement is by itself, but also how it ties into our intention of why we are doing a movement in the first place and what outcome we are looking for.

If this does not resonate with you and you instead insist on not investigating and questioning what you read and hear (including the material in this book), nothing or no one will stop you from doing so. In fact, you will join the majority of fitness "experts," personal trainers, and bodyworkers out there. What I can tell you though, is that this approach of not questioning is the single most effective way of putting a halt to learning, knowledge, insight, and deepening your own understanding of anything in your life. To simply accept something because someone said or wrote it or because a whole industry is practicing it a certain way, will do very little for your own understanding. Important to note here is that I am not saying that everything someone says or writes is necessarily wrong—it might very well be correct—however, I am talking about our own understanding and learning.

What do you think would have happened if Andreas Vesalius had not questioned the understanding of anatomy at the time? I think the answer would be, "not much." Briefly, Vesalius laid the foundation of present day anatomy. Unfortunately, it took quite some time until Vesalius in his turn was questioned, and as a result of this, we now find ourselves with "specialists" for basically every single major joint or body part you can think of. Again, this shows how simply accepting something as true and no longer questioning it is guaranteed to lead to a very limited understanding, even to the degree that the whole (in this case the body as a whole) is neglected or, even worse, not being recognized due to a lack of knowledge.

So, was everything Vesalius said wrong? No, of course not. At the same time, it was not the complete picture. So, thanks to people like Thomas Myers, Thomas Hanna, Moshe Feldenkrais, Gary Ward, and the authors of this book, to mention a few, our understanding of human anatomy and how the body functions and moves in a living body continues to deepen.

This brings us back to the initial question of what movement is. Is movement only active? Can it also be passive? If we take a moment and focus on the passive aspect of movement, we will come to see that there is so much passive movement continuously taking place in our bodies in each and every moment. Every single breath is a testament to one of these passive movements. Of course, breathing can be an intentional act, but most of the time, throughout our day, it is a passive movement. Some of you might say, "To sit down and simply breathe is not movement!" Again, how do we define movement? What are the results and/or benefits a movement has to have in order to be called "movement?" If movement is about the physical act of moving, then breathing is certainly a movement since the diaphragm, belly, ribcage, and

lungs extend and contract with each inhalation and exhalation. If movement is somehow connected to the physical outcome and/or benefits it delivers then, again, breathing should by all means be classified as movement since it can have a far deeper impact on your overall health than most physical exercises alone. I think this is something a whole community of yogis could attest to.

We can acknowledge that movement includes not only physical exercise or physically moving but also the more subtle aspects of movement like the air entering and leaving our bodies with each inhalation and exhalation, blood flowing through our veins with each heartbeat, each eye blink—movements that might be considered insignificant for a movement practice, but, if we think about it, are the very foundation for any other movement to take place.

Imagine if you could not blink. I would think any other form of so-called movement would be rather hard to perform after a while, not to mention if you could not breathe or if your heart stopped pumping blood throughout your body. That would be the end of physical movement altogether for that individual. This is all to say that when we think about movement, I would like to encourage you to broaden your perspective, to think of the body as a whole, not only the big and grand movements but also the subtle ones, not only the active movements you perform but also the passive ones that are always there in the background, not only when you move your own body but also when you are being moved by someone else or you are moving someone else. I believe that to start to approach movement and movement practices from this perspective can begin to open up a canvas of possibilities in which we can find health and well-being that is not limited to short-term results but can be something that continues to deepen and grow as we continue to explore this possibility.

One example of a well-known and practiced movement that seems very few people question the effectiveness of is the Cat-Cow. In this movement, as you extend the spine and arch the back, you also arch the neck and look up toward the ceiling. This is the "cow." When doing the "cat" you instead round the back and neck and look down toward the floor. You continue by

alternating these two movements. Now, is the structure of this movement what the body naturally does? In fact, the neck naturally does the opposite of what the spine does. So, if the spine extends, the neck naturally flexes, and if the spine flexes, the neck extends. The easiest way to get an experience of this is to lie on your back with the soles of the feet against the ground with bent knees. With the heels close to the hips, inhale and arch the back, then exhale and flatten the back, gently pressing it toward the floor. If you pay attention to what your head is doing in this movement, you will (hopefully) find that as you inhale and arch the back, your chin comes closer to your chest, which means that your neck is flexing, and as you exhale and flatten the back, your chin moves back toward the ceiling, i.e. the neck is extending.

These movements of the neck and skull happen quite naturally as you arch and flatten the back in a supine position. So, what we see here is that the neck naturally does the opposite of what the spine is doing. So, in the Cat-Cow, why are we not doing this? Is it wrong to extend your neck as your spine extends instead of letting the neck do the opposite? I would say, depending on your intention and purpose of doing the movement, you can do whatever you want. When you have a clear intention and know what you want to achieve, I would say there are no rights and wrongs when it comes to movement. Certainly, there is a map of what the bones, joints, and muscles naturally do in relation to each other in different positions and phases of gait as they move and this, of course, is what we want to achieve, but there is not one single method that is the only way to achieve this. The interest here, at least to me, is that a learning takes place in the nervous system

so that we can achieve long-term results and not only the momentary "feels-so-good" outcomes. So, with the Cat-Cow movement, a more interesting way of asking this question is, "Why are you doing or teaching this movement the way you do?" Only you know the answer to this question, and it is a question worth asking. Are you doing it or asking one of your clients to do it in a particular way because you have a clear intention of what you want to achieve, or are you simply doing it because that is how you were told to do it, or because that is what everyone else does?

To end this section, I would like to challenge you to take your favorite movement and begin to question it. Why? Because when we begin to question why we do things, in this case a movement, we will set out on a journey to explore the full potential of this movement and give it our full attention. I will walk you through, briefly, how you can approach the movement of your choice step-by-step:

Begin by turning off the music. To some this might seem radical, to others not, but so much comes back to simply listening. Our bodies are constantly communicating with us, and if we were to just listen, we could find so much information, from "just" that.

Ask yourself: Why am I doing this movement? What is my intention? What is it that I want to achieve? How does this movement connect with the rest of my body? Is it structured and performed in a way that is in line with what the bones, joints, and muscles naturally do, not only in one but in all three planes of movement? I am not saying that you have to have the answers to all of these questions right now, but just begin to approach your movement practice in this way. I promise you, it will be so much more interesting, fun, and rewarding to move and teach movement when you can answer or even begin to ask the questions above.

Before beginning to move, simply check in with how your body feels, right now, and then move on to doing a body scan. If you do the body scan standing up, you might want to check whether you carry more weight on one of your legs, where the pressure is under your feet, how your pelvis, spine, ribcage, and neck move in all three planes, going through them one by one and then together. If you do it lying on your back, you can check where the pressure is with regards to your heels, pelvis, shoulder blades, and skull—is there more pressure to the right or left with regards to each of these? How does your lower body feel between the right and left sides? What about the upper body? How does your body feel as a whole? What would your body print look like? These are just a few, brief suggestions. (If you are not familiar with this kind of practice, I am aware that the above is limited in scope and I would suggest that you seek out a workshop with, for example, a practitioner in the Hanna Somatics and/or Feldenkrais methods, an Aspera Education practitioner, or an AiM practitioner with an understanding of how the body moves in three planes of motion.)

When beginning to explore this movement and getting a (new) sense for it, try to connect into the smaller and more subtle movements that are taking place in your body. Ask yourself, "What is happening to my breathing as I do this movement? How was my breathing before doing this movement? How is it afterward? Does my experience shift if I come into the movement on an inhalation or exhalation? Does my experience change if I come out of the movement on an inhalation or exhalation? What happens if I hold my breath during the movement? What happens if I let go of trying to control the breath and simply let it move freely as I perform the movement?" The breath is really the key to many movements, so use, explore, and play with it.

I would also encourage you to experiment how fast you do the movement. What happens if you slow down the whole movement? Do you get a clearer idea of what is happening in the different phases of the movement

as well as the different parts of your body and in your body as a whole? How do things change if you come out of the movement at half the speed you went into the movement with? How are things different if you do the movement as slowly as you can? What happens if you do the movement as fast as you can?

You can continue to explore the range of the movement. What happens if you go to your end range of the movement? Does something change if you instead go to 75%, 50%, 25%, or 10% of your maximum range? How is the movement different at each percentage of this range? What happens if instead of physically performing the movement you simply imagine it? Are you still experiencing any physical effects or other benefits? Many of us have a tendency of often working at the end range of our capacity of a movement. If, as well as practicing at the end range we begin to explore the other ranges we can move in, we might find that there is so much more to learn. Moving small, down to the point where no physical movement is perceivable from the outside, can sometimes have the biggest effect. So do not neglect this possibility.

Next, gaze. What happens if you keep your eyes open during the movement? Does something change if you close your eyes? How does it affect the movement if you let your gaze move in the opposite direction of where your head goes? What about if you let your gaze move up and down while your head turns right and left, or let the gaze move right and left as the head moves up and down. By simply playing with the gaze, you will find that you can make a movement easier or (a lot) harder without changing anything else.

These are just a few suggestions of how you can explore and play with a movement of your choice. With this type of outline, we can begin to look at movement as an exploration instead of "this is how you should do it." Just from this simple outline above, you might begin to get a sense of the potential of a single movement. A movement lesson does not need to consist of ten different movements. Instead, we can simply take one movement and ask ourselves, "In how many different ways can I perform this movement?" You might find yourself "busy" with this one movement for some time to come.

Movement and the mind

"All the qualities of your natural mind—peace, openness, relaxation, and clarity—are present in your mind just as it is. You don't have to do anything different. You don't have to shift or change your awareness. All you have to do while observing your mind is to recognize the qualities it already has."[1]
—Mingyur Rinpoche

The inquiry of the previous section leads us to what I would like to discuss here, namely, what is the connection between movement and the mind? This is a question that could have many possible answers depending on one's focus. I will choose to focus more broadly and open up for an inquiry that will hopefully lead to more questioning and perhaps spark an interest in you as a reader to explore this topic more on your own. However, in order to begin to answer this question, we first need to look into what the mind is—just as we did in the previous section with movement.

It seems that for a lot of people, especially Westerners, when asked what the mind is, we immediately think of the brain and point to our head. However, is the mind limited to the brain? If you ask an Asian person the same question, many seem to point to or put their hand to their heart. A Western person pointing toward their head seems to suggest that the mind has more of an intellectual quality, whereas an Asian person pointing to their heart seems to suggest that the mind has more of an emotional or intuitive quality. It is interesting how, dependent on our view or how we define something, we come to think of or relate to this same thing.

So, what is the right answer? Is the mind more intellectual or emotional in nature? Or, do we have to limit it to one of these two? It seems to me that taking these two examples above of Westerners and Easterners has not so much to do with what the mind is, but more to do with our own perception of it. Often, we seem to take our own perception for the truth instead of realizing that it is simply just that, a perception, and might actually not be in line with how things actually are.

One of my teachers over the years, Tsoknyi Rinpoche, a Tibetan Buddhist lama, covered this nicely in the phrase, "Real, but not true."[2] He is pointing to the crucial point that, for example, an emotional response we can have with regard to a memory is real from the perspective of experience—your personal experience—but the conditions and circumstances on which this experience is based are not necessarily true. So, with regards to the mind and body connection, an example of this could be someone who is experiencing pain in a limb that was amputated. We cannot tell this person that their pain is not real, because this man or woman has a first person experience of pain taking place in their soma; however, we can say that at the same time as the pain they experience is real, it is not true because their limb is not there anymore.

Todd Hargrove, in his book *A Guide to Better Movement*, gives an example of an experiment illustrating the body/mind connection,

"One hand is placed on a table, the other out of sight behind a screen. A rubber hand is placed to the side of the hand that remains in sight. The subject's hand behind the screen is then stroked with a brush while he watches the rubber hand stroked in the same way. Pretty soon he will get an uncanny sense that the rubber hand is part of his body, and he will even flinch when it is threatened.

This means the brain takes 'ownership' of the rubber hand. Even more interesting is the brain also disowns or 'neglects' the hand that is out of sight. We know this because the hand behind the screen actually gets colder, the result of a change in blood flow."[3]

Again, can we deny that the physical sensation the person in this experiment experiences when a rubber hand, which is placed in front of him or her, is being stroked? No, we cannot. The experience is real. However, is it true? We could say that while the experience is real, at the same time, it is not true because while the subject is experiencing this sensation, there are simply no neurological connections to the subject's body whatsoever. However, through the power of the mind of the subject, the subject's body "believes" that the rubber hand belongs to him or herself and therefore begins to relate to it as their own.

Coming back to the mind and our perception of it, it is interesting to me to see that so many of us Westerners think of the mind as something intellectual, and at the same time many of us seem to be quite out of touch with our bodies. We live in a society where science, academia, intellectual knowledge, etc. are highly praised. Having this emphasis has led to a lot of insight and breakthroughs in different areas but, at the same time, we need to acknowledge the need to see the interconnectedness between the body and mind.

Ken Robinson, in his Ted Talk, "Do Schools Kill Creativity," says,

> *"I like university professors, but you know, we shouldn't hold them up as the high-water mark of all human achievement. They're just a form of life, another form of life. But they're rather curious, and I say this out of affection for them. There's something curious about professors in my experience—not all of them, but typically, they live in their heads. They live up there, and slightly to one side. They're disembodied, you know, in a kind of literal way. They look upon their body as a form of transport for their heads. Don't they? It's a way of getting their head to meetings. If you want real evidence of out-of-body experiences, get yourself along to a residential conference of senior academics, and pop into the discotheque on the final night. And there, you will see it. Grown men and women writhing uncontrollably, off the beat. Waiting until it ends so they can go home and write a paper about it."[4]*

Ken Robinson portrays this in a humorous way, but there is so much truth in what he says. We treat our body as a "thing" separate from us that does not do what we want it to do. I often hear people who come to see me talking about their bodies as this dysfunctional "thing" that does not do what it is supposed to do. However, do we listen to our bodies? Our bodies communicate with us all the time. But instead of listening to our bodies and acting upon what it is telling us, we force it into doing what our heads tell us it should do. Like sitting on a chair for eight hours a day, walking around in shoes that were produced to please aesthetics and the fashion industry and not the functionality of the foot. (Do you get a sensation of relief taking off your shoes when you step inside your door, coming home? Your body is communicating with you.) Ken Robinson goes on to say,

> *"Something strikes you when you move to America and travel around the world: Every education system on earth has the same hierarchy of subjects. Every one. It doesn't matter where you go. You'd think it would be otherwise, but it isn't. At the top are mathematics and languages, then the humanities, and at the bottom are the arts. Everywhere on earth. And in pretty much every system too, there's a hierarchy within the arts. Art and music are normally given a higher status in schools than drama and dance. There isn't an education system on the planet that teaches dance everyday to children the way we teach them mathematics. Why? Why not? I think this is rather important. I think math is very important, but so is dance. Children dance all the time if they're allowed to, we all do. We all have bodies, don't we? Did I miss something? Truthfully, what happens is, as children grow up, we start to educate them progressively from the waist up. And then we focus on their heads. And slightly to one side."[5]*

So, what is the solution to this? The solutions can, of course, be many, but one approach could simply be to begin to move more frequently. A number of studies have been made over the past few years showing how exercise heightens the level and length of concentration in schoolchildren and thereby also improves their grades. Now, I am not saying or suggesting that the goal of movement should be to increase children's grades in school or help them to be able to concentrate longer so they can take in more information. I am simply saying it is an interesting example showing the connectedness between the body and the mind.

As adults, I believe it is time for us to drop our view of exercise and training and begin to think of it as movement. As Linus Johansson, the author of this book, has pointed out: there is no substitute for moving well. There are, however, substitutes for strength and endurance. If you need to lift a piano, you can use some tools to help you with that—there is no need to use brute force to get that piano off the ground. The same goes for endurance. For most of us, being able to run 20 km straight or get into a sprint for several minutes is no longer part of our day-to-day necessities. There are very few life-or-death threats now for which you need to be able to do just that. However, for moving well, there is no substitute.

In a study from 2012,[6] a group of researchers made an experiment where they had people do a sit-stand-sit test (for an example of this test see note[7]) and found that adults between the ages of 51 and 80 who moved better had a better chance of an increased lifespan, whereas those with less mobility ran the risk of passing away at an earlier age.

Of course, to move more does not solve the issue at its root, but I believe it can be a step for the majority of people to begin to see that the body and mind are interconnected, and the more we work with and train our bodies and our minds the readier we become to see that the body and mind are truly inseparable.

Just as we need to move and train our bodies—or rather retain, or for many of us adults regain, movement and mobility in our bodies—we also need to take care of our minds. To only focus on the physical or only focus on the mental too often leads to an imbalance where we are not functioning at our best.

Lastly, I believe we should ask ourselves why we are doing this in the first place. Why do I want to move well? Why do I want to be strong? Why do I want to have phenomenal endurance? Why do I want to work with my mind? When we look closely at this question of "why" I think we will come to see (at least I have) that at the end of the day, it all comes down to us wanting to be happy. That is something that we all share. No one wakes up in the morning thinking, "Today I should do everything I can to feel miserable." Even those of us who inflict harm upon ourselves do so in order to be happy. Of course, the issue here is whether we find the real causes of happiness or whether we engage in causes that lead to more pain and suffering for others and ourselves.

Having a healthy body will not in itself bring us happiness. You can have an amazingly mobile, strong, and agile body and be miserable. I think this is something that most of us will agree upon. True and lasting happiness is something that is found within and accessed primarily through the mind. That being said, to be strong, mobile, and physically pain-free can really support and, again, help you understand the inseparability of the body and the mind. Mingyur Rinpoche—who is the most important person in my life for working with and understanding my own mind—reflects upon his own experience regarding the connection between the mind and happiness,

> *"In hindsight, I can see that the basis of my anxiety lay in the fact that I hadn't truly recognized the real nature of my mind. I had a basic intellectual understanding, but not the kind of direct experience that would have enabled me to see that whatever terror or discomfort I felt was a product of my own mind, and that the unshakable basis of serenity, confidence, and happiness was closer to me than my own eyes."*[8]

To end with, regardless of where we come from, if we are a man or a woman, young or old, moving well or not so well, we all want to be happy. It is my belief that through seeing the inseparability of the body and mind, to treat, respect, and be grateful to our bodies, as well as recognizing, applying, and seeing the power of awareness, we can become balanced, healthy human beings who embrace our being as a whole without neglecting any aspects of it. I think the most amazing outcome of being a healthy human being who, through awareness, recognizes these good qualities within oneself, will be that one also begin to see these same qualities within others, which will probably lead to the greatest movement of all.

References

1. Mingyur Rinpoche, Swanson E; *The Joy of Living: Unlocking the Secret and Science of Happiness.* New York: Harmony Books, 2007, p. 98.

2. Tsoknyi Rinpoche, Swanson E; *Open Heart, Open Mind: A Guide to Inner Transformation.* New York: Harmony Books, 2012, p. 10.

3. Hargrove T; *A Guide to Better Movement: The Science and Practice of Moving with More Skill and Less Pain.* Seattle: Better Movement, 2014, p. 109.

4. Robinson K; "Transcript of 'Do Schools Kill Creativity?'" (TED: Ideas worth spreading, February 2006). Available at <https://www.ted.

com/talks/ken_robinson_says_schools_kill_creativity/transcript?referrer=playlist-the_most_popular_talks_of_all> (accessed June 6, 2018>.

5. Ibid.

6. Barbosa Barreto de Brito L et al.; Ability to sit and rise from the floor as a predictor of all-cause mortality. *European Journal of Preventive Cardiology*, 13 December 2012. Available at <http://journals.sagepub.com/doi/abs/10.1177/2047487312471759> (accessed June 9, 2018).

7. Wilson B; Simple sitting test predicts how long you'll live. *Discover Magazine*, 8 September 2014. Available at <http://discovermagazine.com/2013/nov/05-sit-down> (accessed June 9, 2018).

8. Mingyur Rinpoche, Swanson E; *The Joy of Living: Unlocking the Secret and Science of Happiness.* New York: Harmony Books, 2007, p. 13.

Lena Björnsdotter

Lena Björnsdotter is a personal trainer, fitness instructor, and educator. Lena has a BA in sports science and is a certified Anatomy in Motion practitioner. She works with clients and instructs a range of different fitness classes. Lena creates workshops and events to inspire and presents education via Aspera education to other therapists and trainers. Lena is one of the founders of the movement concept Three Planes of Motion, 3PM®.

CHAPTER 22

What is movement?

Lena Björnsdotter

"The least movement is of importance to all nature. The entire ocean is affected by a pebble."
—Blaise Pascal

The answer to the question "what is movement" could be as simple as the dictionary definition. The Cambridge Dictionary refers to movement as, "a change of position." For me, that's too simplistic a view, when it can be answered in so many ways.

From my poetic subjective perspective, I would describe movement as the start of something. That something is the journey through everything. It's the reachable change, the unreachable nothingness, but still a continuous motion that never stops. You are movement. The world is movement. Externally and internally everything moves. Bodies, bones, muscles, thoughts, feelings, heartbeats, roads, politics, religion, and nature, all the way down to the wings of a bumblebee, are movement. I work with movement every day, bodies moving, rolling, rewarding conversations that move, moments and smiles that bounce, all the chains of connecting, moving and connecting again in endlessness.

You are movement when moving forward, through a fast run, with propelling extension and powerful stances. You are movement when walking by the beach, and your eyes take in the scenery.

You are movement when sleeping, when your lungs expand and deflate, when your blood flows and your stress level lowers, while hormone and digestive systems restore. You are movement when sitting down in stillness meditating, following your breath, feeling the shirt tremble with each breath. You are movement when you are not experiencing it, when standing still with your feet on the ground and the joined forces of gravity, earth and you collide. You are movement when there is no movement.

Nothing is ever still. Something is always moving.

Anatomy from the old perspective

To be able to answer the question, "what is movement" I need to separate anatomy into an old perspective and a new perspective. From the old perspective, we need to grasp the exact word of anatomy, because it reflects history. The word "anatomy" comes from the Greek words *ana* meaning "up" and *tome* meaning "a cutting". Anatomy is the identification and description of the structures of living things; it is the scientific study of the body and how its parts and bits are arranged.

The anatomy of the human body has historically been a subject of curiosity and study. All the way from the Greek philosophers to Humorism, the theory of the four humors as a way to map the body. From Renaissance polymath Leonardo da Vinci, who dissected corpses illegally to understand the human body and could recreate amazing art after his discoveries, to English students engaging in grave robbing, to Frederik Ruysch who found a way to preserve corpses via embalming techniques. In the eighteenth century, major steps were taken in anatomy, and dissection was made mandatory for medical students. Studies of anatomy today often employ X-ray and MRI and CT scanning techniques, rather than dissection.

The old divided perspective of anatomy has lasted until modern times. When I did my bachelor's degree in sports science ten years ago, I too learned this descriptive anatomy. We spent a great deal of time studying the Latin names for every bone and muscle, and all the origins and insertions. Sure, we also did palpation exercises on each other in the study groups, and analyses of movement, but it was hard to understand the body in movement from a one-dimensional perspective. It was sometimes also related to mathematics or biomechanics, which made movement hard to grasp and less "living."

While we spent many weeks studying the bits and pieces of the human body and employed mathematics to deepen our knowledge, we spent rather less time discovering anatomy in movement, and no time at all studying the layers of the connective tissue. We spent weeks on the anatomy of the upper body and the legs, but very little engaging in the anatomy or function of the feet. It was as though they were disconnected from the rest of the body—not that I knew the importance of connective tissue and the feet then, but it's interesting in relation to where I am today.

To briefly summarize: the history of anatomy shows that the body's structure has traditionally been described, divided, and separated into different pieces and parts, such as bones, muscles, tendons, arteries, veins, and organs, via dissection. You could say that bodies were taken apart to try understanding a functional and a continuous system.

Movement from the old perspective

"Life requires movement."
—Aristotle

According to the Cambridge Dictionary, movement is a physical change of position, a moving of your body or an object from position A to position B. According to the old perspective of anatomy, a movement derives from the skeleton being like a building with levers, creating the actual movement via concentric muscle contraction over isolated structures. The skeleton is almost looked upon as a hanger for all its parts, which "hang" onto it, and as a protection for the inner organs. In my study literature I can read that the muscles have three major functions: to produce movement, maintain posture, and generate heat.

Movement can be a change of position such as an arm going from flexion to extension or vice versa. It is the route of a body part, caused by a muscular contraction. It is a muscle's way through shortening or lengthening, or the other way around. It could be a biceps curl with a dumbbell or a lying leg press in a training machine.

Historically there has always been an interest in movement as an aspect of being healthy and strong. Luigi Galvani was the first professor that created movement and twitching in the thigh muscle of a dead frog via electric metals; he later developed an understanding of how the nerves control muscles.

His nephew Giovanni Aldini took his work a little bit further, applying electro-stimulating techniques to a deceased human body and stimulating movement of dead limbs to an audience. The results of the experiments seem to be linked to the understanding of muscle function and the directions the muscles create, such as flexion, extension, abduction, adduction, and so on. From my point of view, these ideas appear to have been transferred into isolating specific muscles and body parts in modern training machines and training exercises.

If you test muscle function in order to fully understand movement using dead humans lying on an examination table, you might have some issues. Firstly, they are dead. Secondly, they are not tested standing upright with the force of gravity affecting them.

To conclude, I can see a relationship between the old anatomy's descriptive perspective with all the bits and parts, deceased people receiving small electric stimulants to demonstrate muscle function, and today's training machines isolating specific body parts.

If we go all the way back to prehistoric times, we can surmise that movement was performed in play, in hunting, and during battle. As far as we know, exercises and training first started being organized in ancient Greece, in gymnasiums (sporting and educational facilities open only to adult males, whose name came from the Greek word *gymnós*, meaning "naked", as training was carried out naked). The training consisted of gymnastics, running, wrestling, and spear throwing with an underlying purpose of preparing for battle. The Greeks met the movement needs of what they needed to accomplish in daily life by being strong, fast, and having endurance.

Since then the relative importance allotted to training and movement has been variable. It was not until the late nineteenth century that Dudley A. Sargent from Harvard University developed and tested the first tensile machines and the first leg press machine. He was the first innovator to believe that every muscle could be developed with the right sort of training for that specific muscle. In 1901, the first bodybuilder, Eugen Sandow, arranged his own bodybuilding competition, at the Royal Albert Hall in London. He based the idea of the "perfect physique" on the Greek ideal, and himself resembled a Greek or Roman statue. The training method he developed is still used today.

The old perspective on anatomy and movement is done in a concentric manner, in one anatomical plane, and usually divided into targeting specific muscles or muscle groups, isolating and overloading to get muscle growth without a connective thought.

Since the old perspective of movement reaches all the way up to today, training also has a more modern perspective as well. Movement can originate from basic movements, such as lift, push, press, pull, jump, bend, run, rotate, etc. However, scientific studies recommend that the exercises should be performed with strict technique, straight lines, and inside the box. Training exercises are still concentric and often done in one anatomical plane.

In "free weight training," it sounds as though you ought to be free when moving with weights. Even when training in whole chains and more complex movement patterns over multiple joints, like for example a squat, we still tend to think about these movements in a very one-dimensional way. Does today's movement perspective consider the body as a whole body in movement? Does it connect and act systemically? Do training exercises done in one anatomical plane for 15 repetitions and followed by another anatomical plane for 15 repetitions count as a continuous movement?

My story

I have practiced running, but not in an effortless way. Short legs, a lack of patience, and boredom with the feeling of endlessly moving in the same way have made running a real challenge for me. However, nowadays I avoid running for entirely different reasons. About ten years ago I decided to become a runner. I had a feeling it was going to get easier if I did not give up. I aimed to pass 10 k, and I managed it pretty quickly. I ran for a whole year, 10 k or more, several times a week. I felt like a runner, until I "broke."

I suspected periostitis and decided to take time off from all the training for a month: no running, no dancing, nothing at all. After two months it was no better, in fact it had become worse. The pain was severe and it hurt to walk, so finally I booked a doctor's appointment.

The doctor took one look at my feet, told me that I had flat feet and suggested that I should rest from training.

With the doctor's guidance, I saw a physiotherapist. The physiotherapist did help me—after yelling "this is a catastrophe"—confirming that I was "a pronator" after putting me in front of a mirror and pointing at my medial malleoli that were heading for the ground. That was the first time that I had seen my feet from ground level, in a mirror, and I remember thinking, they do not look like I imagine them to look. Still, I was so thankful, the physiotherapist had a solution, and I was given hope.

I did not realize the yearlong struggle I had waiting for me. Since I craved and longed for movement and training to feel alive, I really put a lot of effort into the physiotherapist's training program so I could get back as soon as possible to a physical body. Three or four times a week I performed all stabilizing movements for ankle, knee, and hip in my new shoes with pronation support and insoles to help stop my foot from pronating. Every movement was locked, stabilized, and isolated to create a new center for my feet and body. When I took walks my feet bled due to a new position and the sharp edges from the insoles. Eventually, it started to feel better and pain-free days came more regularly. After a year I could go on regular walks with my shoes and insoles. One thing I did notice though, was the feel of fragile, unstable, and collapsed feet whenever I took my shoes off. The other thing I noticed was that I had recurrent pain in my lower back, on the left side. It felt like my leg was longer on that side and that my gait had changed during this period.

At this time I was studying for my bachelor's degree in sports medicine, and I could not help my thoughts wandering. How come that you strengthen your "core" when you have lower back problems? How come that you train and strengthen your back muscles and stretch your pectoralis muscles when you want a better posture? Usually, when it comes to pain or rehabilitation, the rule is to strengthen the opposite area of your body as an equation of opposite thinking. So my next thought was, is it really opposite exercises or is it actually a created movement?

How come that you train your whole body when it comes to pain relief, but not your feet? During my rehabilitation process, I was strengthening everything *but* my feet, and my feet were locked in an elevated forced supinated position. I had never heard about strengthening my feet during this process, just stopping any movement from happening. Why not try to create the opposite movement?

Even though I have never thought of myself as a runner I have always felt like a dancer, so dancing was something that was looming on the horizon but felt impossible now with my thick shoes and my insoles. In dance class I was the only one with shoes on, so after a few months I did the forbidden, I took them off against my physiotherapist's recommendation. I was amazed when I realized that dancing barefoot didn't hurt. My thoughts started wandering again: how come just standing and walking around in the kitchen at home caused so much pain in my ankles and feet, but dancing barefoot did not? Was it the change of direction, was it dancing on my forefeet more than the whole foot, was it the dynamic bouncing upload through the arches?

The questions I had formed a curiosity for human movement and gave me a feeling that something was missing. The missing part was a new perspective, how it all connects. Slowly I experimented at home with exercises for my feet to create movement, but it was not until I read about Anatomy in Motion by Gary Ward and the "Finding center" courses in London that I realized that I could get some keys regarding how it all connects.

A continuous system and the new perspective

"A continuous system, a system whose inputs and outputs are capable of changing at any instant of time."

Medical research is often described in a one-dimensional way and unfortunately all the statistical analysis comes via linear thinking. Whenever I am reading scientific articles, following media, talking to colleagues and clients, it all comes from a linear perspective. Everyone states linear answers to systemic questions. Like there's a start and an end? It is either this or that way, without putting it into a context with a systemic approach. Can you understand the body or movement with linear thinking, with linear answers, when everything is cyclic, systemic, and continuous?

The old perspective on anatomy and the old perspective on movement might not just be an old perspective; it is a linear perception that seems recurrent everywhere in society, not least in medicine, physiotherapy, nutrition, and training. This text is supposed to be about movement, but to be able to talk about how I work with movement I need to emphasize that it is not just the view on anatomy and movement that is linear, it is everywhere and prevents development; since everything is moving, it is all connected.

In my personal, and simplified, experience I find that in medicine, the common thing is to look at the symptoms of an illness and treat the symptoms with medicine in the hope of getting closer to the source of the illness. In the worst-case scenario, medicines create side effects with new symptoms and new illnesses. Very often this is a result of linear thinking and linear treatment.

In physiotherapy, you often get indicators that there is a need for movement in rehabilitation, but the contradiction to that is to hinder the actual movement. The linear perception and the linear methods used when a position or movement causes pain could be to hinder movement and stabilize muscles to reach a painless state. On my journey of rehabilitation I was forbidden to take another step before my insoles hindering pronation were made by the orthopedist and my shoes with pronation support were bought, also as a result of linear perception. If one is unlucky, the side effect of hindering movement is that the movement needs to go somewhere, let's say to the next connective segment or higher up in the body chain. If you try to stop movement in one place, movement will happen where it's allowed.

It has always been a mission for me to try to understand everything I take on. It concerns me when no one seems to wonder *why* things happen, like nothing relates to anything. Without claiming that modern medicine and modern physiotherapy are not experts in their area, I really urge everyone in medicine, physiotherapy, nutrition, and training to ask the question *why*— there is a need of a wider terrain and a new systemic perspective.

My working method

When I had done rehabilitation for over a year, with a sense that there was something crucial missing, and had been experimenting with movement by myself, I took the Anatomy in Motion course in London in 2011. That was the starting point that later developed into Aspera Education. During these years Aspera Education has matured and stands for a new perspective, a systemic approach and sense for the continuous system that my colleagues and I have striven for.

I have worked with my feet, creating movement, doing exercises, and nowadays I don't use insoles at all or shoes with pronation support. And if I feel the slightest thing, I know exactly what to do, to make a difference. During these years of exploring movement, I have one true conclusion and one clear insight. Movement is always the answer, and if I had known what I know today about movement, I could have rehabilitated myself in weeks instead of over a year.

A few years ago a colleague came up to me, asking me to help her. She said her right hip hurt so much that she could barely walk. We had been working a long time together. I had seen her daily when walking toward, beside and behind her, and like a snapshot I got a sharp mental picture of her gait. I got clear inner images of how she walked, her posture and how her feet looked when changing shoes in the dressing room. It is like I had registered bodily details, all the limitations in her movement patterns and the potential of possibilities that had passed by me daily. I had already seen, processed and understood it. I had not only seen the continuous system and the solutions. I also knew the answer to the question why. After many years of studying movement, seeing and understanding how the body works had become a part of me.

External perspective

When I work with clients, I end up in the sensory range of movement. My working method can be divided into an external perspective and an internal perspective. From the external perspective, I mainly register through my eyes and my hands. During the years I have done a significant number of visual assessments, functional testing, and gait analysis. I have unconsciously registered clients' and non-clients' posture, feet, gait and movement patterns. It is automatic—even if someone just moves in the corner of my eye, they leave an impression.

I always start a consultation with a visual assessment or functional testing, depending on the client and the client's goals. Through static visual assessment, I externally register the relationship between, for example, feet, pelvis, and ribcage, together with functional testing. I acknowledge the limitations and the possibilities, the movement and the non-movement. In the visual assessment, I see the whole body and where the force of movement has been planted in the body. In the functional testing I get the answers to what I have seen in the visual assessment and create exercises according to the findings.

In my work method, I create connective movement in line with how the skeleton moves. The understanding of how the feet function gives me clear indications of what is happening in the whole body.

In contrast to the old perspective that comes from an anatomical plane view on movement, I work in three planes of motion, integrating all anatomical planes at the same time from the feet up to the crown. In contrast to the old perspective I don't only work with movement in a concentric manner over isolated structures. I work in an eccentric manner, using fascial recoil to create movement, linking the feet to the crown, and find that the eccentric exercises make movement happen where movement has been lacking, pain problems dissolve, and it is performance enhancing.

When creating movement I have my "extra glasses" on to see the clients actual movement, through position, the center of gravity, acceleration of movement, breathing and down to the tremble of the skin. I use my hands to lead the client and create movement into gentle awareness, with my hands; I also feel if the soft tissue is moving or if it is hindered. When my clients are in pain, I explore movement because there will always be a small space in which they can move without pain. When leading the client, close enough to pain but not interfering with the pain, I can work around it and eventually create small movements that lead to more

significant movements like ripples on the water, and most likely to freedom from pain.

In contrast to the old perspective, I see how movement is connected and related inside the body. I also see the potential of the client and where I can take them. I recognize how the feet move and interact with the rest of the body. When thinking about training, I understand, feel, and sense the actual movement of how the skeleton moves rather than seeing training exercises engaging specific muscles. In contrast to the old perspective, I can see where clients are right now, and what possibilities there are for them to be able to own their movement.

Internal perspective

Movement could also be seen in a more intimate and sensory context that is experienced from the inside. I believe that many people have forgotten, do not want to acknowledge, or do not know how to experience their body. It could be the experience of their own movement patterns and how they respond to all kinds of emotion, such as, for example, anxiety and sadness that has transferred and been planted in their body. The result of that could be an experience of pain, stiffness, and immobility. If working with body awareness in a sensory context and connecting movement you might realize that you can change your whole persona. The changes could be easier breathing, more integrity, stronger speech, being able to lower your stress levels and to be more of you. To be your movement is to own your movement.

If you google the two words "experience movement," the first thing that comes up is from a woman's perspective, fetal movements. Why is it that we talk about experiencing movement only when something is actually moving and changing position inside of you? If you exclude dictionary definitions of movement as a change of position of an object or a person from A to B, what is a sensory experience of movement?

We never talk about movement as a sensation. A sensation of movement could be to feel the slightest movement by being aware. I work a lot with movement awareness, to let the client be aware of the feeling of how the feet connect to the rest of the body. The communication between the client and me could be the client's experience to be aware of muscle contact and the ground through their feet, also to be aware of trembles, changes in tension, side differences, and breathing. I work with small movement changes, including all anatomical planes at the same time; it could be playful movements that challenge the client's movement box, and the sensory result from this could be a resistance linked to control or a fear of pain. But that fear is also a sensory experience that too many keep too close, as an identity. Working with the client and collaborating within this internal perspective makes the resistance loosen up and the client gains access to their movement.

I have noticed a changing view on movement, but also a need for something new to arise. Hopefully, this book could be the starting point for that to happen. Slowly something is happening. Slowly something is moving.

I still believe that the question why needs to be asked more often and a systemic approach needs to be awakened, and not only in the movement field.

Linear thinking, a one-dimensional view and one given answer to a systemic question is about to be historical.

I am. You are. We are. A continuous system.

Gary Ward

Gary is the author of *What The Foot?* and creator of Anatomy in Motion. Gary has spent years observing the human gait cycle and mapping the movement journey of each bone and joint in all three dimensions through a single stride. Each movement is described in his Flow Motion Model˚. Gary teaches his Finding Center course, a six-day immersive experience in the closed chain biomechanics of the human body.

Gary is known for his upside down thinking, thinking differently about anatomy. Turning many conventional ideas on their heads has brought Gary to the forefront of many people's attention and he has become known for helping people to heal where many others fail to do so. Not through some magical technique, but by using human movement patterns derived in his Flow Motion Model to re-educate and restore long lost movement, and critically, he encourages patients to finally take ownership of their own problems instead of relying on someone else to be their fixer …

CHAPTER 23

What is the potential influence of the skeleton's movement on the fascial system?

Gary Ward

I am a proponent of movement. Movement is my thing. For as long as I can remember I have had access to movement, utilized movement to the fullest, and worked to have control over the many parts of my human body from having fun with it, to training it, lifting weights on it, growing it, damaging it, repairing it, working on it and with it, and back to having fun with it again! My biggest driver though was most probably the fear of losing it. I did not want to grow old and not be blessed with mobility. Not satisfied with having it, and not wanting to lose it, drove me to need to understand it, so that I might preserve it. I'm lucky enough to be able to share my findings and my fascination with the subject of movement with many people around the world. Those people come from all manner of bodyworking backgrounds.

Some people use movement and exercise as their bodywork tool of choice, others use their hands manually, and some a mixture of both. All disciplines come from their own school and have their own school of thought. Often, though, it's not what I would call movement. For me, when trying to understand movement, I became engaged in how our human body moves. I soon realized that it is less about how the whole body moves as a unit, rather it concerns how all of the parts move together to form that whole. Those parts, of course, are the bones and the joints. The hard tissue. Attached to this hard tissue are all the muscles and enveloping all of it (as well as coursing through it) is the fascial system.

I am a proponent of movement, so lying someone down on a couch to work on their body to promote their ability to move well has never been my thing (despite this I am fully aware of the value and benefit of this form of bodywork). I say this because I want to quickly move toward an understanding between the reader and me that I am willing to state that the condition of your fascial system is entirely dependent on your bone and joint system having full and unrivalled access to its full potential. If every single joint is fully able to make use of its three-dimensional potential—that is to say fully able to access its planes of movement, be it one, two, or all three—you will find a certain optimal swagger in your body's movement, you'll find that tissue and fascia no longer hold the limitations we are used to seeing in the body, and you will notice a flow in the body unavailable currently to many people.

My interest is wholly in how the skeleton moves and in how it influences tissue. I studied the joints to observe how they move, I then used these joint motions to determine how soft tissue must respond to movement and was able to construct a whole body model of the gait cycle that describes the three-dimensional motion of every joint in the human body through the journey of a single footstep.

How joints move gives clarity to how we walk and helps to reason why we walk the way we do; it also guides us to make sense of why someone might be walking the way they are walking (e.g. with a limp, etc.). When people walk with a limp, it's possible to connect this to the hard tissue being compromised somewhere, likely in relation to or compensating for a previous experience in their body. The gait model I created, which I have called the Flow Motion Model™, describes the journey of each joint through the gait cycle. It describes each and every joint's position at each individual moment in the walking cycle. To then place each joint in position while standing upright creates an exaggerated whole body posture that is representative of a position we access in each moment when walking.

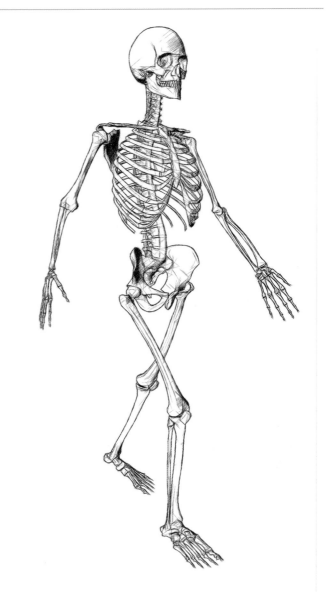

There are 12 such positions in my Flow Motion Model. I call them phases. Gait phases if you will. In each phase, you could say that each joint is working together to create the whole body posture since each joint seems to be able to communicate movement up and down the kinetic chain. Movement X at the foot means I can create movement Y in the pelvis. To limit X is also to limit Y; or a lack of available motion at X means that Y accesses excess movement to make up for the lack at X. Such anatomical interplay is at large through the whole body and in each phase. The patterns seem to be set in stone. The relationships occurring within them, equally so. Having overlaid Tom Myers's Anatomy Trains work to the Flow Motion Model many years ago, I was delighted to see that

I could use Tom's work to make sense of people's muscle tensions when I matched the muscle to how the joint system was accessing its movement. My ultimate fascination came when I discovered that giving correct movement back to the skeleton was in itself sufficient to make changes in the fascial system. One student of Myers's came out of two minutes of closed chain upright skull, ribcage, and pelvis mobilization likening it to the experience of Myers's front and back line work. I bring up the anecdote merely to underline my area of interest, which is how purely the movement of the bones and joints within our skeleton can influence fascia. A model of movement that is unique to the bones and joints and impacts on everything within the body is what makes this work so appealing to a wide base of bodyworkers.

In order to get started, we should consider what movement even is before observing it in both the joint system and the fascial system. I approach this text in the full awareness that movement in the realm of anatomy is more and more a common idea for "modern exercise." Movement is an input into the system, and the thing that movement puts back into the system is … movement. Movement restores movement. It gets things moving again. Movement feeds movement. A treatment, the use of touch, a manipulation or any essence of control placed on a body to help it overcome a complaint, an injury, or whatever reason drives people to see a therapist, is not movement. Many an input is placed into the system—a massage, for instance, mobilizes tissue and lymph, manipulations for the skeleton generate a movement in the receiving human body, and yet the receiving body experiences it passively; many exercise-based approaches still embrace the essence of stability or stabilization, the intention of which is likely to control movement, and yet cannot be described as putting movement into the system since to stabilize a part of the body is to deprive it of its necessary motion.

I looked the word up in search of a definition: Movement—"an act of moving."

I searched again for the word "move" and came up with, "to go in a specified direction or manner; to change position."

I can, of course, easily build this out into the notion of movement being related to exercise; for instance, I could run in a specified direction, in which case I would be moving my body, or I could do a squat or a push-up moving from one body position to another. In both cases I am certainly moving, but then walking in a slovenly manner around a supermarket could also be defined in this sense as movement. Or as the right-handed beer drinker would like to say as he raises his pint glass from A to B, he too is "exercising" (his bicep).

If we were to view, for a moment, the body as a system … again I sought out a definition for "system" which read as follows, "A set of things working together as parts of a mechanism or an interconnecting network. A complex whole." We can definitely consider the human body as a system, especially if we substitute the word mechanism with organism. We have done a wonderful job over the years of investigating all of the parts that make up the human body, and we do know a lot about them. It would appear that information we have gleaned about the human body was collected from non-moving bodies. However, to observe movement of the system in the closed chain—best interpreted as "upright, alive and interacting with external forces"—does change one's interpretation of many things, for instance how muscles work, how far-flung joints appear to interrelate with each other, and how being upright can influence our thinking compared to lying on a treatment table. We have a lot of individual and isolated information about each part of the human body and yet, unfortunately, they do not inform us about what we need to know about movement of the overall organism so we can understand the role of this word "movement" within the system. Movement is challenging to discuss. On the outside everything appears to move. On the inside everything needs to move, but often doesn't move in a way that would best serve the system.

Inside this system, then, as the human either (1) moves from A to B or (2) changes the position of itself to express movement, are multiple interconnected moving parts. One of the challenges is that despite therapists going to great pains to learn as much as they can about the various moving parts in the human body, the people they treat, who rely on the information arising from the fitness and therapy industries, most likely are ignorant of their many moving parts. People go to the gym, go for a run, participate in their chosen sport, passion, profession and are naturally unaware of how each part of their system is moving, and whether it is moving well or not … they happily leave this question to their therapist.

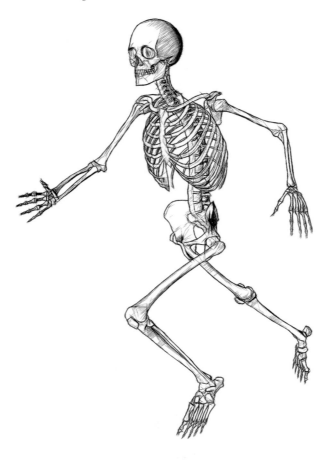

My focus and interest has always been placed in our body's bones and their motion; not just their individual motion but in particular their interconnected and interrelated motion, all of which culminates in the overall motion potential of the body as a complex whole or system.

All movements for all sports require us to move our bones, and therefore it is equally possible to observe the motion of those bones and their joints. First consider that it is possible that there are only a handful of ways that our joints interact together to form whole body movements required to walk. If we consider these as an optimal set of movements and consider each one a pattern of movement, we can then begin to look for these optimal patterns of bone motion in individuals and compare them to the patterns being

used by an individual in any given movement. To do so might create an opportunity to help optimize their performance or make sense of difficulties they may be having in that pattern (due to how they are choosing to access their movement).

So I began to look at the interconnected movement of the individual parts within the human system. I continue to be fascinated by the unbreakable patterns that we use when walking and their correlation to many, many other activities. (When I say unbreakable, we can distort the patterns, but rarely, if ever, without compromise to the system in some limiting or adaptive way.) The movements of the individual parts come together to enable us to access the whole body movements required to access all of the many activities enjoyed by human beings.

Can we enjoy such activities if our moving parts don't do what they should? Yes, of course we can, and very many of us do … it may or may not surprise you that all over the world, the courses I run to teach my interpretation of movement are full of people who describe themselves as "movement people," people who move, most of whom are shocked by how many of their own body parts fail to move as they could, or dare I say, should?

Should? Who says that the body should move in a particular way? Well, it turns out that I do. Mainly because I subscribe to a notion of balance and an optimal way of moving for every joint and every bone in every human being, that is determined by the individual human being. This can be observed by how easily that individual can access each joint motion in each movement pattern. Being able to access all joint motions should create a balanced system.

Optimal: Best or most favorable.

Balance: An even distribution of weight enabling someone or something to remain upright and steady. A situation in which different elements are equal or in the correct proportions.

Both of these definitions, for me, pertain to the human being. So based on the fact that we have two legs, two arms, and a central column running down the body, creating two sides, it makes sense to me that as humans, we should be using both sides of our body evenly if we are to get the most out of our system.

For me, three words/phrases express the value of being able to access both sides of a human's (or a single joint's) movement:

+ effortless
+ conservation of energy
+ efficiency.

What is movement in relation to the joint system?

Movement can be effortless; it can also be effortful. The human body is made up of multiple joints, each with its own unique range. Range in the majority of joints (it may be easiest to think about this in the spine) means that such joints would be best served to have equal access in both directions, either side of a center or midline. It is, I think, generally recognized, that to be optimal, a spine would equally side bend both left and right, to either side of its neutral position, and the same could be said for most joints (barring the knee and elbow since they rest more extended than flexed and not in a centered state). Therefore, it makes sense, does it not, that if a spine can move laterally more left than right, then there would be immediate effort placed upon the system. Why? Each joint is crossed by soft tissue: muscle, fascia, tendon, and ligament. An optimal lateral bend to the right in the spine for instance means that the joint space should open on the left hand side of the spine and close on the right, ideally equally at each vertebra but more often at some more than others.

All over the body, where joints gap or open, muscle, fascia, tendon, and ligament all increase in length or lengthen; where joints close, muscle, fascia, tendon, and ligament all reduce in length or shorten. Lengthened tissues are said to be under tension. I call it the horse's head analogy: you are minding your own business, sitting atop a horse, holding the reins, which attach to its bit, and the horse lowers his head to the ground. The gap between you and the horse's head increases and the tension in the reins increases too. If you don't go with the horse, the effort required in you to maintain the same position is increased. It is no longer an

effortless posture, but an effortful one. It's no different for the human body as the brain demands greater output from the tissues on the left side of the spine than the right.

Quality is synonymous with effort. If all joints open equally in a side bend, this is an effortless side bend. If some joints struggle to open on the left when side bending right, while others open excessively, we still have a side bend right and yet this side bend is marred in quality by the distortion in the bend, thus contributing to more effort and less conservation of energy, as some soft tissue on the left side experiences more lengthening, while others remain short or inactive.

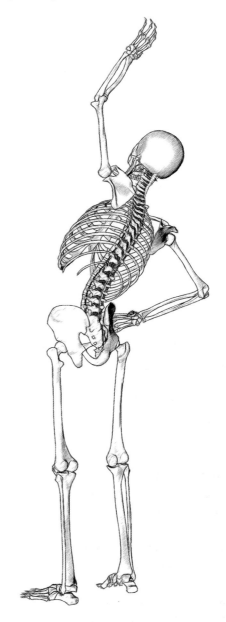

You should be able to feel in your own body that as you side bend to the right, even if only slightly, the weight is drawn into your right foot. As you weight-bear more on one foot than the other, you begin to use that leg more and the other leg less. Is this balanced? No. Does it create an effortless situation for movement in your system? No. Would moving around on such imbalances be conserving of energy? No, it would not.

Now we have to zoom out a little and consider that word posture, because what we are describing at one joint could be happening at many, if not all, joints in the body. More open on one side than the other. If one joint is not balanced at rest, can any of them be? Posture is perceived as how we stand, sit, and present ourselves to the world. Posture is actually the sum overall resting position of each joint in the body and the resting position of each joint is based on the sum potential of its own movement availability.

Some terms you may be aware of when currently discussing or considering posture are: good; bad; forward head; round shouldered; kyphotic; lordotic; anterior tilt; knock kneed; flat foot/high arched. Interestingly, none of these words describe movement. Movement within the system is what I ultimately wanted to define when I first set out on this journey of anatomical investigation.

As a personal trainer in the early 2000s, one of the key things we looked at was posture—not just in terms of the words above, but in terms of the body's overall position in space. The body is a three-dimensional entity. Most of human anatomy focuses exclusively on one dimension of movement and one joint/structure at a time. I was taught to look at it in all three. I wanted to know if a person's pelvis was neutral in all three planes, not just one. This means it would be neither anteriorly nor posteriorly tilted, elevated neither on one side nor on the other, and rotated neither left nor right. Add to that the idea that it would sit evenly over both feet. That's a neutral pelvis. On top of such a level and neutral pelvis, a spine can stand tall and proud. Neither flexed nor extended with no side bend or rotation present in it at all. I didn't come across this very often. I would say that almost every pelvis and all spines had some distortion or other present in their standing, resting posture. This gives me a very clear

indication, not only of what soft tissue in the body is under tension, but also about the potential for each section to experience its desired movement.

Back to movement; it's actually very simple in a being with two sides and a central column. Each structure and each joint should be able to experience all of its available motion in all available directions either side of its midline or center.

A pelvis that is rotated left at rest actually struggles to rotate right when the person is walking around or going about their daily activities. This means that tissues that are involved in the right rotation of the pelvis are lengthened, under tension, and unable to shorten to their full capacity. Such muscles remain full time in an effortful space and begin to fatigue and can become problematic. A pelvis that is excessively anteriorly tilted has the same problem; it likely struggles to posteriorly tilt, and the tissues involved in the posterior tilt of the pelvis are lengthened, under tension, and unable to shorten to their full capacity. Some of you reading this will have a pelvis that is both anteriorly tilted and left rotated. If you do, it's likely that your pelvis is hiked up on the right as well as adding tension to your left adductors and right side abductors. There is a reason for such bold claims.

The shapes of our interconnecting joint surfaces determine how we walk. I do believe there is the perfect way to walk, the ideal gait, and I know that it cannot be taught, it simply has to have your full arsenal of joint motion available. Not in just one dimension of movement, but all three. I became aware that each moment in the gait cycle creates a new shape in the body. Each shape can be described differently by virtue of observing the movement of the bones and their subsequently related joints.

There are, in my opinion, twelve different postures that we should be able to access when we are walking if we are to experience effortless, energy conserving and efficient walking patterns. I have them all defined in my Flow Motion Model. Each one of these 12 postures is described by both the position and direction of movement of each structure and each joint. Naturally each posture is different and thus represents 12 combinations and patterns in which we are readily and willingly able to move our bones and joints. This is where the movement patterns begin to show themselves in the model and are easily traceable across the many sports and activities we do in daily life. Perhaps the sports we play today would be different had our human body's moved and developed differently?

One of the things I noticed early in my observation of the human body when walking is that anybody walking over a force pressure platform would take anywhere between 0.6 and 0.8 seconds to complete a single heel to toe footstep. In that time the human body should (again) experience every single possible joint motion available in both directions (e.g. flexion/extension, adduction/abduction, etc.) and in all three dimensions available to it. Now that is a lot of movement. When people tell me they want to get better at watching gait, I am skeptical. Good luck tracking 217 bones and 300+ joints as someone walks on a treadmill (incidentally, a powered treadmill does distort these natural movements). At best you can pick out the gross movements of the structures, but never the finer, more specific detail. This short time frame is, interestingly, sufficient to govern the body's potential. A pelvis that is rotated left at rest has insufficient time (in the 0.6–0.8 second window) to experience an equal amount of rotation to the opposite right hand side, thereby ensuring the pelvis experiences more left rotation than right rotation in its everyday movement (imagine what this would do to the piriformis, and other hip musculature, for instance). The limitations here are not necessarily at the pelvis, but in the overall makeup of the body and the time it takes the foot to get from heel to toe. Perhaps if we were to influence the timings of the body's overall motion in the gait cycle, such limitations at the pelvis might resolve themselves.

What I invite and educate people to do is to get better at observing how each joint and structure in the body experiences its own individual movement, while paying attention to the interconnections and patterns of movement that the body seamlessly undertakes. Each joint connects above and below. If joint A relates above to joint B and joint B relates above to joint C, then it also makes sense that joint A and C must also interrelate in some way. Carry this out ad infinitum to the relationship between joint A and joint Z.

I'd like to take a step back for a moment. Remember the words we used to describe posture? None of them describe movement. They describe a position in space. This position in space could be anywhere on the movement spectrum. A movement spectrum is an infinite amount of points between one end range and its opposite. If a pelvis is anteriorly tilted, by how much? And what is its impact on the structures above and below? An anteriorly tilted pelvis may well be able to posteriorly tilt, but may not be able to do so beyond its true neutral. This would at least suggest there is some range present, it's just not as big as we would like or covering an appropriate area within the required range. This suggests that soft tissue on one side experiences a shortening and on the opposing side a lengthening, and never its opposite state. Far better for less range to be available but the anterior and posterior tilts are able to cross the midline somewhat—at least opposing muscles would have a chance to lengthen and shorten with more equality, even if not by a desirable amount, and their resting length may be the same.

An anteriorly tilted pelvis could also be stuck in its tilt, unable to tilt more or less than the position it currently occupies; this suggests a stuck-ness with very little requirement for soft tissue involvement. Arguably this could be more of an overall problem for the system. Range of motion is an interesting aspect of our "movement." A joint that rests in the center or in its neutral position has more range than at any other point on its spectrum. The more toward one extreme of the spectrum it rests, the less overall access to range the joint has. The further away from center in either direction results in less and less range being available at the joint. Thus to give range back to a joint is not about stretching a muscle but facilitating the joint to experience neutral (which requires joint A below and C above to also be considered and involved in the experience of a new movement). Movement is a direction. Forward, backward, sideways, or in rotation (to the left or the right).

Movement is symbolized by two things; (bones) moving toward each other or away from each other in their respective planes of motion and the outcome at the joint is simply one of change. Movement can also be considered more globally, such as a shift of weight toward one foot or away from it. Such a change in position heavily influences the position of the bones and interconnecting joints as some open and some close to accommodate the shift in body weight. This is what I call "mass management."

How we manage our mass also pre-defines our movement potential. Our mass can be forward in our forefeet, backward in our heels, over our left foot, over our right foot, and a rotation in our system may even have you experience your mass in one heel and the opposite forefoot. The weight distribution in your feet—your foot pressure—is something you can dial into pretty quickly in your own body and can give you a huge insight into your movement potential. I mean would you choose to squat under a heavily loaded bar knowing that you have nearly 40% of your whole body weight through one heel? What could be the consequences? Well, frankly, if we break it down into three dimensions, we could say that your weight is backward toward the heels; this is usually (but not always) associated with an anteriorly tilted pelvis.

We can say that your weight is more to one side than the other, let's assume right, and is likely associated with a pelvic hike up on that right side encouraging more weight to load into that heel bone. Given that there is so much weight in the right hand side we can also guesstimate that there is very little weight in the opposite forefoot, which suggests that the pelvis is rotated to the left—this is also a logical choice for the brain as to rotate the pelvis to the right would increase the load in the already burdened right heel. So not only are we squatting on a heel that bears the majority of the weight, we are also squatting on a pelvis that is anteriorly tilted, hiked up on the right, and rotated to the left. If you pay attention to your spine, the lumbar spine in particular, there are natural associations that your spine must make in order to accommodate such an adjustment in the pelvis.

Your lumbar spine will be extended on the anterior tilt, side bending to the right on the hiked up pelvis and rotated to the right on the left rotated pelvis so as to maintain a forward facing axis in the upper structures while you stand at rest. Can you work out which part of your body you naturally load up by squatting in this position? Your right side low back and sacroiliac

area, which is extended, side bent right, and rotated right, force closing the upper sacroiliac joint structures each time you squat (other adaptations are of course possible, but very few in fact, and in reality it is these postures that will be right in front of you on display in the patient—just remember to check!)

Your foot pressures indicate what potential posture you are holding, which defines both your movement limitations and potential for movement inside your system which in turn defines what you are physically able to achieve when performing any given movement. The movement efficiency of your whole system is reliant on the movement potential at each joint in the system and is certainly hopeful of an air of balance or equality on both sides. Whether that arises out of your joint motion potential as defined by your foot pressures or out of your foot pressures as defined by your joint motion potential, balance in both is a key ingredient to efficient movement.

Our current thought processes around movement remain based in the usage of soft tissue to alter our posture or a joint's position in space. There is this idea that by actively contracting tissue we can encourage a structure to change its shape and in the long term generate a new posture. I have to disagree, and this is where many philosophies can easily be challenged. Contracting tissue may well move a joint, but will likely not move a joint to its end range, if that end range cannot physically be achieved within the joint. When this is the case, correct movement of the joint in conjunction with relative and appropriate movement above and below (throughout the whole system) will enable that joint to access its full range (in the absence of physical or structural blockages of course) while also creating an environment for the muscle to fully contract. While I am fully aware that it is possible to contract a muscle and move a joint, it does require the joint to have access to its full potential.

What I observed in human movement is that muscles lengthen before they contract, and I have described this as one of my big rules of motion in my book *What The Foot?* That is to say that when we are in motion or movement our skeleton intermittently accesses (or is required to access) different positions throughout the duration of that motion and in each case, as we move

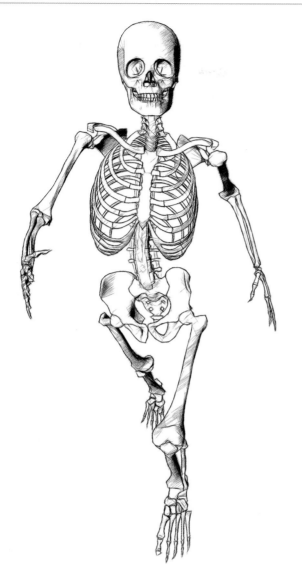

Joint motion is necessary for muscles to have a role. I do not believe that the opposite is true. Now consider all of the stabilizing or contraction work you have been doing to either close joints or keep joints stable and you'll hopefully realize that a large slice of the pie is being overlooked. For that reason I created a second big rule of motion: joints ACT: muscles REACT, which is aimed at helping us to make sense of bone and joint motion when considering movement either as a therapeutic tool or for exercise. If a joint doesn't have mobility, the tissue surrounding it has nothing to do. The tissue around it is not "doing a good job of keeping it stable." The joint is simply movement-less. A joint can lack mobility in any of three states:

1. Closed shut (surrounding tissue would be locked short). A joint that is closed is likely unable to open by much in its 0.6–0.8 second window of opportunity. Joints that are more closed than open start to have reduced range the closer they get toward their end range.
2. Open (surrounding tissue would be long). A joint that is open is likely unable to close by much in its 0.6–0.8 second window of opportunity. As a joint rests closer to its end range, the range available in both directions reduces.
3. Stuck or immobile (surrounding tissue would have nothing to do). A joint can be stuck anywhere along the spectrum of its supposed potential for movement.

our joints, muscles are lengthened by the opening of these joints and the necessary muscles subsequently contract to control the amount of motion allowed at the joint. Muscles contract to control the amount of movement a joint may access—not to promote movement in the joint. Movement in the joint arises out of the joint having access to range based on its resting position in space. Resting in a centered position, it may have equal access to both end ranges. Resting off-center it does not. Thus muscles contract eccentrically to allow a joint to move toward their end range, and concentrically when in its longest state to bring that joint back toward its neutral position. Muscles, in motion, do not contract from a mid-range to short but always from a lengthened range toward short. Thus it takes a joint to open or gap to give a muscle something to do.

Again looking at the vertebral column, which may rest in a lateral flexion to the right. The vertebrae would be open on the left hand side and closed on the right. Tissue on the left side would be long and tissue on the right side short. The tissue on the left hand side is actually working hard to limit further opening of the intervertebral joint, while the tissue on the right has nothing to do, due to lack of stimulus from the joint being closed. It is a common thought that the short tissue is contracted to hold the body in that posture, whereas I suggest that due to the joints being closed, that short muscle has, instead, nothing to do, and will in fact only respond to a lengthening request via an opening of the associated joint(s).

Our temptation may be to stretch the tissues of the right side, or contract the tissues on the left or both.

And yet, it is the movement of the joints that will enable our tissue to have a role to play again. Lateral flexion of the spine, again, is a whole body motion. Being right laterally flexed means a limitation is placed on the neck, the pelvis, the hip joint, the knee, and the foot.

Working all of the joints together in their relevant movement patterns is what will re-educate the nervous system to appreciate what a straight spinal column actually is. As the spine straightens, the tissue on the left side will contract or shorten as the tissue on the right hand side lengthens to allow the joint to gap on the right and close on the left. How we got here is a different conversation but the outcome is the desired goal of movement, to realign our structures, re-educate our tissue and create a more efficient system to move around on (or in). By moving the joints we create a much greater opportunity for soft tissue to fulfill its role of managing our skeleton in movement. I think that joints create movement, while muscles are perfectly set up, instead, to manage the available range and accessible movement within our joints. They can allow too much movement or too little movement, neither of which creates a useful environment for movement of the system, as that would require just the right amount of movement and exchange between our muscles and our joints.

What then is movement in relation to the fascial system?

Time is a factor here. Many people profess that working on the fascial system and all the nerves and proprioceptors held within is what we need to do to reorganize our posture. I have a different view, and I hope it is simply taken as another side of the coin that we can add into our fascial conversation. The science around fascia is incredible and in fact mind blowing. To listen to the likes of Tom Myers, Phil Beech, and Robert Schleip discuss fascia and the magic within is nothing short of jaw dropping and awe inspiring. But what if something is being overlooked? And that something, I believe, is why I was asked to write this piece, because that something is "movement." I do not think any of the above fascial proponents would disagree. When I first started teaching Anatomy in Motion, I wanted something to support the Flow

Motion Model and its joint motion interpretations and the only thing I found was Tom Myers's Anatomy Trains. Fascia overlays, surrounds, and fuses through everything, both soft tissue and hard as the periosteum of the bone (outer layer) is infused within this huge fascial container surrounding the human skeleton.

The big question for me remains: does fascia move bone or does bone move fascia? I would be foolish to say that one does and one does not. It is such a difficult conversation to engage in, most simply because the two are so close together in their role in the human body that it is arguably impossible to separate the two. And yet, there already exists this one conversation that fascia can be worked with in order to reorganize a human skeleton, that muscle can contract and move bones and joints and the primary focus is usually on the soft tissue governing the bones and joints. I want to have the "other" conversation, the one I am most inspired by, having worked with countless patients over the years and have observed changes in fascial tissue in crazy short time frames by going at it "bone first."

Simply put, if you can re-educate the brain to re-experience an old (forgotten) movement pattern that it has long been unable to access, it is possible for the skeletal structure to reorganize itself. In doing so the fascial system must instantaneously adapt as well. Whether or not you subscribe to chains or trains of fascial lines that run from head to toe, or if you are a proponent of the notion that anybody can wield a scalpel and come up with their own interpretation of a fascial line, one thing is for sure: fascia interconnects everything! Given this fact, it is easy to see how it's possible to improve the quality of one's fascial system and the whole body will adapt positively. In exactly the same breath, it should be possible to improve the quality of a person's joint motion and the whole fascial system will adapt positively. If a spine straightens out from a simple movement pattern, i.e. you stand up taller after than you were before, then your whole fascial system must have changed to allow this adjustment in whole body position.

Movement is repetitive. Over millennia, human beings have ground their bones into the shapes they are today through the desire to be upright and in motion. Walking on two feet, eyes on the horizon. Every step we take is nigh on identical to the previous footstep. It was the same yesterday and most likely the same last month—go back far enough and it may have been the same for years. Unless something happened to change it. Usually what happened is negative. An injury, a trauma, an accident, a disease, a surgery … in each case something happened to the physical body that caused it to adapt. If you injure your ankle, it's not just your ankle that is affected; it is your whole body. If you injure your shoulder, it is not your shoulder that is affected; it is your whole body. As your whole body changes, so does the way you walk, which in turn alters your footstep, your stride, and the timing differential between each leg. This impacts on your three-dimensional posture, opening some joints, closing others, lengthening some tissues and shortening others. Your posture changes. Movement is repetitive. The new learned movement patterns settle in and the way you walk is different now to how it was before.

The good news is that your ankle does not cause you any trouble any more, and yet, the irony is that the way you now walk is likely to cause you trouble as new stresses are placed on other parts of the body. If you really damaged that ankle, how safe would you feel putting your whole body weight on it? If you haven't done that in the few weeks since you damaged the ankle, it's likely that you'll never do it again! Well you might put weight on your foot but are you actually able to fully commit to it? The way you manage your mass changes to adapt to injury, your whole skeletal system adjusts to find a new way of standing, walking, and moving. Your fascial system is also along for the ride. The timings of each footstep change and directly influence bone and joint motion in the whole body. Bones are now finding that in this new repetitive way of moving, they go more left than they do right. Some prefer to go forward and not back.

Joints are now opening more on the left than the right, etc. Muscle tissues that span the joints to manage them have increased demand and are placed under more effort as the body conserves less and less energy. Muscles being a part of the fascial system means we can now observe the changes in these "fascial lines" quite easily as we observe the bone motion and test the joint motion potential. Can the joint move equally to both sides of neutral in all three dimensions? If the answer is no, we can describe it as imbalance and can observe

tightness, stiffness, laxity, length, over/under activity (insert your own words) in the fascial system.

We could say that the fascia has tension due to being weak; we could also suggest that the fascia has tension due to the bones opening. I also like to add that the role of the muscle in the fascial system is to control the joint as it opens, therefore it begins to contract against the opening of the joint; successfully done this would close the joint again, but in the case of the adaptations experienced in the skeleton, the joint cannot close (as this would compromise the strategy in place to protect the injury), so the muscle finds itself in a permanent lengthened state. Overactive, attempting to shorten without any success, this is ultimately what makes it weak, tight, and sometimes painful. The fascial system is managing the movement and the position of the skeleton in space, both at rest (posture) and in motion (gait/activity).

Given that I have determined 12 postures in my Flow Motion Model, each posture, in theory, ought to be accessible by all human bodies (but due to life, are generally not); this also suggests 12 different shapes the fascial system ought to be able to achieve. As I've just mentioned, whatever position you put your skeleton into, your muscle and fascial system comes along for the ride. It goes where the bones and joints do—which makes sense since all of the attachments are onto the bones with the skeleton. If a person is unable to access some or all of these 12 postures evenly on both sides then the fascial system will naturally follow the bones and serve to minimize the opening of joints and movements away from neutral.

Only when the skeleton can move from neutral toward its end ranges, back toward neutral and beyond, out to its opposite end ranges and back again, can the fascial system achieve the holy grail of balance, which would be synonymous to the joint system as the two are virtually inseparable. An example of human movement at the pelvis: in an ideal world, you would hike (lateral tilt) your pelvis up on the left when you bear weight on the front left foot and you would hike your pelvis on the right when you bear weight on the front right foot. If you do this equally on both sides then muscles such as abductors and adductors would have a fairly even role to play on both sides in each footstep. What tends to happen is that somebody invariably hikes too much on one side and given the 0.6–0.8 second rule I mentioned earlier, will be unable to equally hike on the other, often just managing to breech or attain a level pelvis when the other side should be hiking up. If I hike on the left when my left foot is forward and only achieve a level pelvis when my right foot is forward, you should notice that the left abductors and right side adductors now have a much greater experience of lengthening than do the equivalent muscles on the opposite side. The right side adductors perpetually experience more length than their left side counterparts—this is often the case with regular groin pulls for instance. There would be very little benefit from treating the tissue of this lengthened groin, when what we really need to happen is for the person to begin hiking up on the right side of the pelvis to match the movement of the left, and to do this naturally with every step they take. When this happens, new length will be experienced in the left groin as the pelvis hikes up on the right and the right groin will be able to shorten reducing the stress/threat on the tissue. If this person was hiking up left due to being unwilling to weight bear on the right ankle after previous damage, for example, then it may not be the pelvic hike that is your focus of attention, but the reason for not choosing to bear weight on the right leg in the first place—the damaged ankle. Treating that ankle (even though the injury was, say, nine years ago) may provide the long-term solution for the person's groin. If the person is comfortable bearing weight on the right ankle, that may be all it takes for the person to begin hiking up on that side again and placing new and improved demands on all of the tissue.

In order to now build this out into a whole body posture, we have to ask the question, "How do I keep my body stacked up over the pelvis in the event of a pelvic hike and with a pronating foot?"

When you hike the pelvis up on the right hand side, for the purposes of mass management, you want the body's center of mass to sit firmly within the weight bearing foot's base of support. This is the essence of balance, when all the centers line up. Thus, below the hiking pelvis the front leg will be bent, adopting a valgus position, dorsiflexing the ankle and flexing at the hip. Above the pelvis the spine will be both side bent and rotated toward the front leg, with a counter side bend

and rotation at the neck in order to keep the eyes on the horizon. This describes the whole body position in the frontal and rotational planes at this moment in time. If the pelvis doesn't hike up, the spine will be affected above as will the hip and knee be affected below. Now consider how much fascial tissue must also be distorted as a result of the change in movement from effortless to effortful. Our goal is surely not only to consider the role that fascia plays in human function but also the role that effective joint motion plays in our ability to walk and move effortlessly.

To understand the impact of movement on the fascial system, it is worth taking on the following consideration. If the manual work we are doing on people's bodies is able to create an environment for the human nervous system to get back to walking and moving in a more efficient way, then the manual work is achieving what it set out to do. Here is the biggest problem I notice: we provide movement exercises that are either not whole body or are not specific enough in the movement requirement. Also, we still have a therapy set that will crack something or offer a treatment that does not affect the whole system. In this case, to crack the lower spine to enable the hip to hike more, in essence freeing it up, may seem like a good outcome until the person walks out the door refusing to unconsciously bear weight on that damaged ankle.

The patient is already walking their problem back into their own system … and many therapists over the years have found clients walking back in with the same problem. So we have to rethink bigger, look wider and deeper, and take a much deeper context into our therapy sessions. That context is whole body movement. How is the person moving around? What movements has the body adopted to protect something in the system? What movements feel safe and what movements feel threatening to the nervous system? Fascia, soft tissue and muscles all contribute to movement, and they are all, in terms of movement, responsible for managing the motion of the bones and joints. The bones and joints are engineered (by us!) to flow in a certain way.

My investigation into human movement has highlighted this in the 12 patterns we adopt as we move through a single footstep and those patterns create a wonderful opportunity to assess the whole body in context and in relation to its history, so that we do not end up treating the complaint, but are able to get closer and closer to the cause. Movement is neither an integrated system nor an isolated system … movement of the whole (body) is wholly reliant on the movement of its parts (bones/joints). We have whole body movements readily available in the gym and therapy field right now, and yet they are inconsiderate of the movement potential at each and every part; and we have isolated processes taking place in the gym and therapy field right now, focusing on individual joints and structures, and yet they too are inconsiderate of the movement of the whole body. It is time to fuse these processes.

Movement is everything. If you choose to follow my line of thinking, movement is also capable of being optimized. Everything is moving as it should, or nothing is able to move as it could. And in this space we find the human being, moving around their problems while creating more in the process. When everything is moving as it should and all joints have their full three dimensional potential, then every muscle has full potential to lengthen and shorten in all three dimensions, every nerve has full potential to slide and communication around the body is unimpeded, every vein and artery is free to pump unrestricted, all lymph channels are free to return the flow back out of the system, every organ is mobilized by the movement of the torso, and the breath is free to inhale and exhale without compromise. A system that moves is a system that can heal itself … it simply requires permission to express it while therapists focus on removing the blockages and limitations that the movement of our life stories has placed upon it, so the system may begin to heal itself. Perhaps movement in all its complexity is simply a mechanism for healing … above all, the healing environment is created by being able to access full joint motion potential, which gives the muscles all they need to manage. The question is no longer does the skeleton influence the fascial system. Moreover it's time to ask the question how do we accurately assess the motion potential within the skeleton?

Jerry Hesch

Jerry Hesch is a Colorado licensed physical therapist with a bachelor's of science in physical therapy from University of New Mexico, and a master's of health science from the University of Indianapolis. He completed a doctorate in physical therapy from AT Still University.

Dr. Hesch has developed the Hesch Method over more than 35 years of clinical practice. The Hesch Method is a distinct and gentler approach than the joint adjustment model and has described numerous patterns of motion dysfunction that exist in patterns throughout the body. The Hesch Method treats from a whole-body approach rather than focusing only on where it hurts. In the lumbopelvic area alone, Dr. Hesch has identified more than a dozen faulty motion patterns not yet described in the literature and developed treatment methods. He has reinterpreted the mechanics of traumatic birth named pubic symphysis diastasis (or pubic joint or symphyseal diastasis) and has developed unique treatment. He has also developed a unique approach to treating the atlantoaxial joint.

Dr. Hesch has taught more than 100 seminars since 1985, instructing clinicians in application of the Hesch Method to assess and treat the pelvis, sacroiliac joint, and lumbar spine. He has also developed learning materials on his advanced body of work, which allows the clinician to apply manual therapy using a whole-body approach. He is currently involved in teaching and writing, and is also accepting patients with complex chronic pain.

CHAPTER 24

Treating micromotion hypomobility of the atlantoaxial joint in patients with whiplash injury

Jerry Hesch

Introduction

A pattern of compression and hyperextension at the upper cervical spine, especially at the atlantoaxial (C1–C2) joint, is frequently encountered via manual evaluation by this author in an outpatient physical therapy practice. This pattern is under-represented in the literature and public domain based on searches on www.PubMed.gov, www.Google.com and www. YouTube.com utilizing key words "atlas, atlantoaxial, extension, hyperextension, flexion, compression, traction, distraction" (date of search December 31, 2018).

However, many resources are readily encountered when the term "rotation atlas" is utilized. This includes evaluation and treatment of rotational asymmetry of the atlas. Rotation is a relevant movement of the atlas and is well represented in web searches whereas distraction and extension hypomobility of the atlas is apparently under-diagnosed and therefore undertreated. This author encounters traction restriction with the upper cervical spine in mild extension frequently in the population with whiplash and cervicogenic headache. This chapter focuses on evaluation and treatment of the upper cervical spine for extension distraction hypomobility.

Articular shape of the occipitoatlantal and atlantoaxial joint

A brief review of the articular anatomy of the occipitoatlantal joint (OAJ) and atlantoaxial joint (AAJ) will be presented (figures 24.1–5). For a detailed review see Bogduk and Mercer,[1] Dalton,[2] and Neumann.[3] Bogduk and Mercer serve as the primary source for the following review.

The OAJ and AAJ lie directly beneath the mastoid processes in the anterior aspect of the spinal column whereas the C2–C7 facet joints lie in the posterior portion (figure 24.1). This distinction is important when applying traction to isolate the upper cervical spine.

The first cervical vertebra, named the atlas (figures 24.4, 24.5), is a ring-like structure designed to transmit forces from the head to the neck. It has two lateral concave facets that are shaped somewhat like a peanut having two connected lobes. These articulate with the convex tubercles of the occiput (figures 24.2, 24.3). The atlas moves in concert with the occiput. The posterior aspect of the anterior arch of the atlas has cartilage and articulates with the anterior aspect of the dens (aka odontoid) process of C2 (figures 24.4, 24.5). There are two lateral facets in which the inferior aspect of the lateral mass of the atlas articulates with the axis. These facet joints are plane-like (figures 24.3, 24.4); however, the cartilage is convex on each surface.

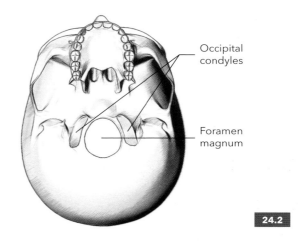

24.2

Convex occipital condyles that articulate with concave facets of the atlas.

Occipital condyles

Foramen magnum

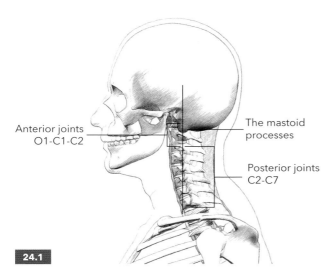

Anterior joints O1-C1-C2

The mastoid processes

Posterior joints C2-C7

24.1

The occipitoatlantal joint is medial to the mastoid processes whereas the atlantoaxial joint is directly beneath the mastoid process. In contrast, the facet joints of the mid and lower cervical spine lie posterior to the mastoid processes.

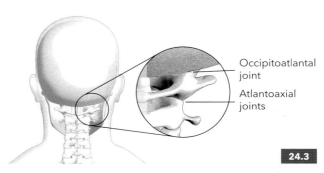

Occipitoatlantal joint

Atlantoaxial joints

24.3

Occipitoatlantal and atlantoaxial joints.

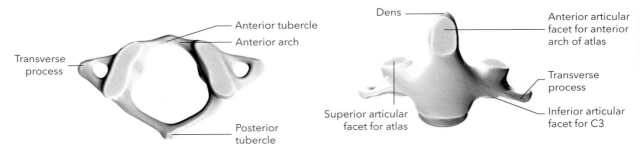

Atlas (C1): superior view

- Transverse process
- Anterior tubercle
- Anterior arch
- Posterior tubercle

Axis (C2): anterior view

- Dens
- Anterior articular facet for anterior arch of atlas
- Transverse process
- Inferior articular facet for C3
- Superior articular facet for atlas

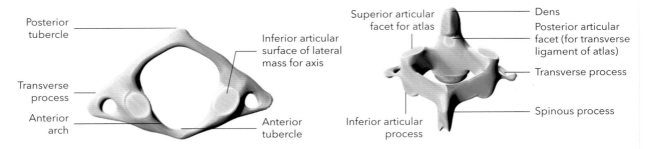

Atlas (C1): inferior view

- Posterior tubercle
- Transverse process
- Anterior arch
- Inferior articular surface of lateral mass for axis
- Anterior tubercle

Axis (C2): posterosuperior view

- Superior articular facet for atlas
- Dens
- Posterior articular facet (for transverse ligament of atlas)
- Transverse process
- Spinous process
- Inferior articular process

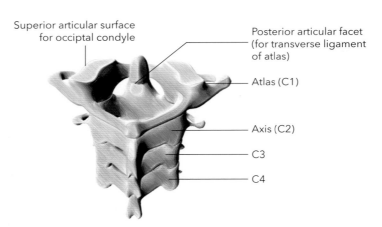

Upper cervical vertebrae, assembled: posterosuperior view

- Superior articular surface for occiptal condyle
- Posterior articular facet (for transverse ligament of atlas)
- Atlas (C1)
- Axis (C2)
- C3
- C4

24.4

Articular anatomy of the atlas and axis.

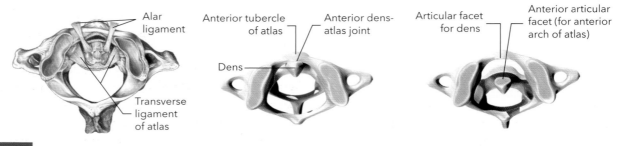

- Alar ligament
- Transverse ligament of atlas
- Anterior tubercle of atlas
- Anterior dens-atlas joint
- Dens
- Articular facet for dens
- Anterior articular facet (for anterior arch of atlas)

24.5

Ligaments of the atlantoaxial joints.

Macromotion and micromotion

Active flexion in the upper cervical spine is under conscious control and is a macromotion/gross motion. It can also be described of as osteokinematic or "observable movement or observable effect of bone movement." When the upper cervical spine moves in forward bending/flexion a rather small amount of flexion motion first occurs in the OAJ and then flexion motion occurs in the C1–C2 articulation and then continues downward. Micromotions are very small movements, which may accompany gross physiological motion but can also be elicited passively by an external force and cannot be isolated with conscious intent. There is a micromotion of distraction that occurs in the posterior aspects of these articulations while there is a micromotion of compression that occurs in the anterior aspects of these joints with cervical flexion. The other micromotion that accompanies these motions is posterior glide and roll occurring with flexion. Rolling and spin are also described as arthrokinematic motion. Roll and spin will not be addressed because these are movements that cannot be isolated with external forces. Arthrokinematic motions are a necessary part of macromotion physiologic movement. Arthrokinematic motions occur within the joint surfaces and has the following aspects according to Mulligan:[4]

+ Unobservable articular accessory motion between adjacent joint surfaces such as roll, segmental glide, and spin.
+ These accessory motions take place with all active and passive movements and are necessary for full, pain free range of motion.
+ Arthrokinematic motion cannot occur independently or voluntarily and if restricted, can limit active physiological movement.
+ Arthrokinematic motion occurs anywhere along the range of motion.

Like arthrokinematic motions, joint play is a micromotion that is not observed and is not under conscious control, thus cannot be isolated with muscle contraction. Joint play can only be evaluated passively. The various types of joint play include distraction, glide in extremities and unisegmental glide in the spine, and overpressure such as at the extremes of joint flexion, extension, and rotation, etc. Joint play is utilized to evaluate micromotion mobility, or hypomobility. Joint play can only be evaluated when the available slack has been taken out of the joint via glide, traction, or overpressure. This chapter will focus on traction of the upper cervical spine performed with the neck in neutral and in 10° of flexion and 10° of extension to optimize isolation of the AAJ. The term traction end feel relates to two mobility states with no gray areas in-between; mobile or hypomobile. It is the nature of this joint to present with either extreme and this property makes for lucid evaluation.

Upper cervical traction end-feel is a unique type of joint play and is a normal physiologic motion when tested at 10° of flexion and extension. This motion is not subtle. The motion is readily observable as is the distinct lack of motion when present. With specific manual contact on the temporoparietal region, distraction is isolated to the upper cervical joints. Following injury this joint play can become hypomobile/blocked and requires passive evaluation and passive treatment, which will be addressed later in the chapter. The term hypomobile in this context is the same as blocked mobility. Restricted mobility at the OAJ is rare whereas restriction at the AAJ is frequently encountered, such as in patients who have had a whiplash injury. Clearing the OA joint and isolating the AAJ is described later in the text. Additional information on principles of joint structure and function can be found.[2,3,4,5]

In order to isolate traction to the AAJ, contact must be made on the temporoparietal region (figures 24.6, 24.7). When AAJ traction hypomobility is encountered in neutral (figure 24.7) and in 10° of extension (figure 24.9), the atlas is presumed to be stuck at the end range of physiologic extension. This author intentionally avoids the subluxation explanatory model. Unfortunately, there is a paucity of published literature on the topic of compressed, hyperextended AAJ including evaluation and treatment per pubmed.gov literature search using the key words "atlas, atlantoaxial, flexion, extension, traction, distraction, traction, compression, hypomobility" (search date November 27, 2018). There is, however, biomechanical literature indicating that flexion at the AAJ is 8° and extension is 10°,[6] and 11.5° flexion and 10.9° extension, 38.9° of unilateral rotation and 6.7° of lateral bending according to Panjabi et al.[7] To review: the atlas forms three articulations with the axis. When there is motion at the lateral atlantoaxial facet joints there is also movement at the anterior articulation of the dens and atlas. With rotation of the atlas there is up to 3 mm of vertical glide at the anterior joint.[8] Bogduk and Mercer[1] (p. 176) also report upward motion of the atlas on the dens based on joint shape, "The odontoid process is curved slightly posteriorly. This shape allows the anterior arch of the atlas to slide upward and slightly backward, thereby allowing the atlas to extend." According to Neumann[3] (p. 281), flexion at the AAJ consists of superior distraction of the spinous process of the atlas and downward pivoting at the lateral facets and at the anterior dens-atlas articulation. Extension induces the opposite. Motion in the upper cervical spine has been objectively measured.

However, paradoxical motion of the atlas has been elaborated upon by Bogduk and Mercer[1] (p. 177): flexion of the head can induce flexion or extension, and extension can induce either flexion or extension of the atlas. This is based on anatomical variation and difference in the location of weight bearing of occiput on atlas. The biconvex shape of the lateral facets suggests that weight bearing anterior or posterior to the apex would predict opposite motion coupling. This may partially explain why traction isolating the axis can be limited in slight flexion or in slight extension after injury such as whiplash. Nonetheless, the most common presentation is one in which traction is hypomobile when tested in neutral and in extension.

Contraindications to evaluation and treatment of the atlantoaxial joint

The same contraindications to upper cervical manipulation apply to passive joint testing and treatment of the upper cervical spine and has been described and is abbreviated below.[9] The reader is encouraged to review the detailed elaboration from the World Health Organization Guidelines.

Inflammatory conditions, such as rheumatoid arthritis, seronegative spondyloarthropathies, demineralization or ligamentous laxity with anatomical subluxation or dislocation, represent an absolute contraindication to joint manipulation in anatomical regions of involvement. Other contraindications include:

1. anomalies such as dens hypoplasia, unstable os odontoideum, etc. This includes developmental anomalies such as Down syndrome, etc.
2. acute fracture
3. spinal cord tumor
4. acute infection such as osteomyelitis, septic discitis, and tuberculosis of the spine
5. meningeal tumor
6. hematomas, whether spinal cord or intracanalicular
7. malignancy of the spine
8. frank disc herniation with accompanying signs of progressive neurological deficit
9. basilar invagination of the upper cervical spine
10. Arnold–Chiari malformation of the upper cervical spine
11. dislocation of a vertebra
12. aggressive types of benign tumors, such as an aneurysmal bone cyst, giant cell tumor, osteoblastoma or osteoid osteoma
13. internal fixation/stabilization devices
14. neoplastic disease of muscle or other soft tissue
15. positive Kernig's or Lhermitte's signs
16. congenital, generalized hypermobility
17. signs or patterns of instability
18. syringomyelia
19. hydrocephalus of unknown etiology
20. diastematomyelia
21. cauda equina syndrome

Mobility testing of the upper cervical spine

Prior to treating the AAJ, the OAJ is evaluated and treated if restriction is encountered. Treatment of the OAJ is beyond the scope of this chapter though there are numerous texts and online resources. The following tests are performed bilaterally in supine in order to relatively unweight the joints and reduce the effects of gravity.

Less force is required when mobility testing is performed in supine as opposed in sitting. The induced motions are very small, and the clinician should attempt to minimize any motion occurring into the mid or lower cervical spine. These tests should be performed very gently.

+ With thumb pad on the anterior surface of the mastoid, an anterior to posterior spring test is performed to evaluated rotation.
+ The pad of the index finger is placed beneath the mastoid process and a superior spring test is imparted to evaluate side bending joint play.
+ Open hand digital or palmar grip of the occiput bilaterally is used with a posterior to anterior lift.
+ Open hand digital or palmar grip of the occiput bilaterally is used with left and right side-glide.

The following tests are used to isolate the AAJ and are performed bilaterally in supine. They can be repeated at C3.

+ The lateral tip of the transverse process is located just below the mastoid processes. Left and right side-glide mobility is evaluated with the pad of the index finger.
+ The undersurface of the transverse processes is palpated bilaterally with the pad of the index finger. Lift up one side and then the other to induce rotation.
+ Palpate the prominent midline spinous process of C2. It is easily located beneath the occiput whereas C1 spinous process is not as prominent and is deeper. The head must be flexed slightly to expose it. Contact with the pad of the index finger and apply an anterior glide force.
+ Use digital or palmar contact bilaterally at the temporoparietal (figure 24.6) region just above and anterior to the mastoid, avoiding contact with the occiput. Test with the neck in neutral, 10° of passive flexion and 10° of passive extension (figures 24.7–24.9). Apply traction to take up the slack. If you are unable to take up the slack the joint is hypomobile and requires treatment. If you can take up the slack to a natural stop, but there is no additional

movement with the spring test, the joint requires treatment. Note that this test may be positive in the presence of a restricted OAJ in which the special tests for the OAJ would also be positive.

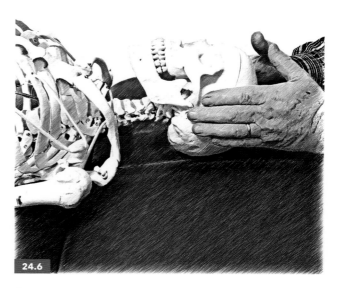

24.6

Contact on the temporoparietal region above and anterior to the mastoid processes will isolate traction force to the upper cervical spine.

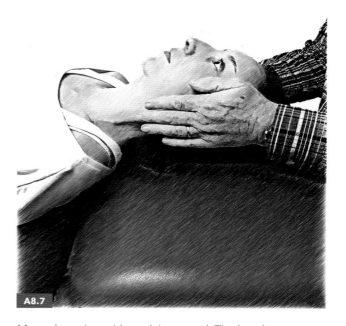

A8.7

Manual traction with neck in neutral. The head is supported by the table. Note the contact on the temporoparietal region.

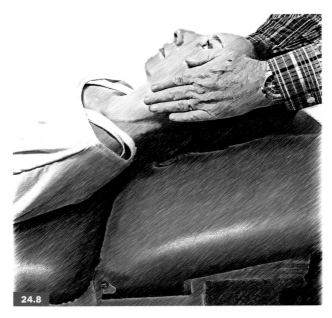

Manual traction with head in 10° of flexion. The head can be supported by an adjustable headpiece or with a low pillow. Note the contact on the temporoparietal region.

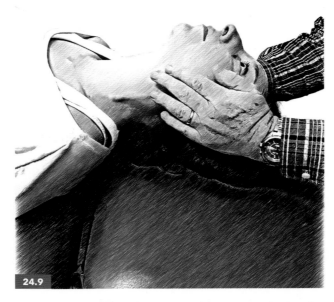

Manual traction with neck in 10° of extension. The head can be supported by an adjustable headpiece or with the head off the table resting on a pillow in the lap of the clinician. Note the contact on the temporoparietal region.

Hyperextension compression of the upper cervical spine

The presumed mechanism of injury involves a passive hyperextension force to the upper cervical spine, such as via a fall, motor vehicle accident, or repetitive trauma.

The most common presentation of upper cervical hypomobility is hyperextension of the AAJ in which traction performed in neutral (figure 24.7) and in 10° of extension (figure 24.9) is hypomobile, whereas traction performed in varying degrees of flexion is of normal mobility. Testing at 10° of flexion (figure 24.8) and extension allows relative isolation of the AAJ and the traction takes up the slack in the OA joint further isolating the AAJ. The temporoparietal manual contact isolates motion to the upper cervical joints. A traction force is applied gently and slowly until the slack is taken out and motion has ceased. This requires approximately 10–15 lb (4.5–6.8 kg) of tension depending on body morphology. In order to develop a felt sense of this force, one can practice by attaching a cervical traction halter onto a fixed fish scale. Alternatively, one could practice on a 10 lb/4.5 kg sack of flour. Note that the average force applied in order to perceive taking up the slack via traction will be slightly greater than the weight of the head due to friction. After that, practice on several asymptomatic individuals should yield an appreciation of normal mobility. Testing the patient population will yield an appreciation of hypomobility.

Taking up the slack occurs in both the AAJ and in the OAJ. In testing, once the slack is taken up (10–12 lb/4.5–5.4 kg) an additional traction force is induced using the same amount of force as was used to take up the slack (additional 10–12 lb/4.5–5.4 kg). The second part of the test evaluates force transmission through the AAJ. The hallmark of passive force translating through the upper cervical into the rest of the body in a non-restricted upper cervical spine is observable movement through the body as distal as in the feet. If the AAJ is hypomobile, movement will not be perceived and it will not be possible to take up the slack. This is not a subtle phenomenon, as a moderate increase in force will still fail to induce mobility. It seems reasonable that the passive testing isolates upper cervical motion because the force is very mild in contrast to studies that recommend 10% of body weight as ideal,[10] and up to 50% of practitioner's body weight[11] for performing traction to the mid and lower cervical spine. Furthermore, the average human head only weighs 9.9–11 lb/4.5–5 kg,[12] based on cadaveric study in which the neck was sectioned at C3. That such a small force can induce movement throughout the whole body in a normal population may initially appear to be counterintuitive; however, with a little bit of experience the reader should be able to appreciate the validity and utility thereof.

Treatment for a hypomobile atlantoaxial joint

Treatment is always performed close to the motion barrier but only in a direction that does allow traction mobility. If traction is hypomobile in neutral (figure 24.7) and in extension (figure 24.9) it is treated in flexion (figure 24.8). If traction is hypomobile in neutral and flexion it is treated in extension. Restricted traction in extension and in neutral is the most common presentation and is treated with manual traction applied at 10° flexion (figure 24.9), propped on a pillow or using an adjustable headpiece, with a specific temporoparietal hand-hold (figure 24.6) used to isolate force to the anterior portion of the upper cervical spine. The force is held for five minutes. The amount of force just matches the available movement of the head and neck and is appropriately described as "very gentle" and averages 10–12 lb/5.4 kg. It is important to mention that the upper two joints (OA and atlantoaxial) are anterior to the facet joints of the mid and lower cervical spine (C3–C7). The AAJ and OA joints are inferior to the mastoid process. Therefore, the hand contact must be very specific in order to avoid force application behind the mastoid. A palmar contact is applied to the temporoparietal region and the direction of pull is toward the vertex. This manual hold is in contrast with traditional manual cervical traction in which traction is applied via contact on the occiput. This important concept of contact above and anterior to the mastoid for upper cervical isolation has not been encountered in the literature and appears to be underappreciated in the clinical domain.

After treatment, traction mobility is retested in the prior restricted positions of neutral and 10° of extension. Only one visit for manual intervention is typical for restoring normal passive mobility. The client is also given isolated upper cervical exercises and self-traction to be performed while lying in a neck cradle named Doctor Riter's Real-Ease® (real-ease.com, Torrance, CA). The neck cradle allows easy isolation of upper cervical movement. A two-inch diameter rolled towel under the mid cervical spine is an inexpensive alternative. These exercises and self-traction can be viewed on www.YouTube.com

by searching, "Hesch Upper Cervical Exercises"). The exercises and self-mobilization are also detailed in the case study below.

Treatment bullet points

+ Inform the patient that they are to report any adverse or unusual response to testing and treatment, such as discomfort, light-headedness, dizziness, visual blurring, tingling or numbness, etc. Inform them that the procedure must be discontinued if these should occur.
+ Determine if hypomobile traction test occurs in 10° of flexion or 10° of extension.
+ If there is traction hypomobility in extension, passively position the head in 10° of flexion.
+ If there is traction hypomobility in flexion, passively position the head in 10° of extension.
+ Contact the head with bilateral palmar contact at the temporoparietal region. This is just above and anterior to the mastoid processes. Avoid occipital contact.
+ Take up the slack by imparting 10–12 lb/5.4 kg to a natural stop.
+ Maintain traction force for five minutes.
+ Retest traction mobility in the position in which it was hypomobile.
+ If mobility is restored proceed to instruct in self-management as described above and in the case study.
+ Schedule follow up visit to reevaluate and review self-management.

Case study

A 66-year-old massage therapist was treated once for hypomobility of the atlas with lack of traction joint play mobility in neutral and in 10° of extension. This is an interesting case study because the client had received twice-weekly chiropractic and osteopathic adjustments for 30 years with an estimated out of pocket expenditure exceeding $50,000.00. The chiropractic treatment was specifically directed at the atlas. She reported that benefit of treatment was very short-term.

The client presented with a lack of upper cervical distraction in neutral and in 10° of extension but had free mobility in 10° of flexion. In flexion the traction force translated through the body and foot motion was observed, thus the traction force was not blocked. She was treated with traction applied with palmar contact on the temporoparietal region bilaterally at 10° of flexion sustained for five minutes (figure 24.8). The amount of traction force applied equaled the amount needed to take up the slack. She responded very positively with a feeling of lightness and better alignment in her head and neck and freer cervical mobility. Passive testing revealed resolution of the hypomobility.

She was instructed in a home exercise program utilizing Doctor Riter's Real-Ease neck cradle which isolates free and easy motion to the upper cervical spine. Very small movements were encouraged in order to isolate the upper cervical spine. These consisted of manual vertical distraction for two minutes once weekly, 30 reps of the following exercises twice weekly: posterior glide coupled with flexion, chin tucks, right and left rotation, right and left side-bending, right and left side-glide. She was also taught to combine sustained cervical spine retraction with repeated chin tucks (upper cervical flexion). Craniocervical retraction isolates upper cervical flexion while extending the mid and lower cervical spine according to Neumann[3] (p. 282). The exercises with the neck cradle and self-traction are available on www. YouTube.com using the search terms, "Hesch Upper Cervical Exercises".

The client was seen twice and did not require passive treatment on the second visit. She provided the following feedback three weeks later:

I have had severe neck issues involving my upper cervical vertebrae since the age of 24 (42 years ago), when I had a bicycling accident. It was initially necessary for me to get chiropractic treatment every week, and sometimes twice a week, just to function. My symptoms were severe occipital headaches, tightened jaw muscles, and vagus nerve involvement that caused heart racing, trouble regulating breathing and swallowing, and basically being stuck in sympathetic nervous system state! As the years went by, I saw many, many different chiropractors and osteopaths! When I first saw you, I was still seeing someone for adjustments at least every other week, if not every week. After you did your traction technique on my upper cervical spine, the relief

was amazing! My skull felt completely comfortable on top of my spine! I watched your video on self-treatment and have been doing it on myself each time I start to feel the headache or internal shakiness or racing heart. It works like magic! Thank you so much for this, I love being able to self-treat. It will save me a lot of time and money!

This is an atypical case study given the symptoms described as being vagally driven. The typical response to upper cervical traction is improved upper cervical traction joint play, greater ease of gross cervical motion, a sense of improved posture and reduction of cervicogenic headache.

Conclusion

Flexion and extension of the AAJ is a normal physiologic movement and can become restricted when coupled with compression in patients who have sustained cervical spine trauma. The manual contact is very specific in order to evaluate distraction of the AAJ. The literature and general body of knowledge has very limited information on this presentation. Treatment is very simple and direct and the same is true for self-management. A typical response to treatment is normalization of all directions of passive mobility of the atlas, with reduction of pain and a sense of optimized cervical posture along with greater ease in active cervical motion. This treatment model is based on very brief intervention and then instructing the patient in self-management. A case study illustrates these concepts. That the traction technique isolates the AAJ is presently only a theoretical construct. Additional research is needed to determine if this is valid and research on patient populations are needed to evaluate the validity, and utility of this technique.

References

1. Bogduk N, Mercer S; Biomechanics of the cervical spine. I: Normal kinematics. *Clinical Biomechanics* (Bristol, Avon) 2000, 15(9):633–648.
2. Dalton D; The vertebral column. In Levangie PK, Norkin CC, Lewek MD; *Joint Structure and Function: A Comprehensive Analysis*, 6th edn. Philadelphia: FA Davis Company, 2011, pp. 154–157.

3. Neumann D; Axial skeleton: osteology and arthrology. In: Neumann D; *Kinesiology of the Musculoskeletal System: Foundations for Rehabilitation, 3rd edn.* St Louis, MO: Mosby Elsevier, 2016, pp. 262–267, 277–286.

4. Mulligan E; Principles of Joint Mobilization. Power Point presentation. https://www.physio-pedia.com/images/c/c0/Principles_of_Joint_Mobilization.pdf (accessed November 26, 2018).

5. Threlkeld J; Basic structure and function of joints. In: Neumann D; *Kinesiology of the Musculoskeletal System: Foundations for Rehabilitation, 3rd edn.* St Louis, MO: Mosby Elsevier, 2016, pp. 25–40.

6. Abernethy J; Upper Cervical. Upper Cervical Spine Orthopedic Residency Lecture. Scottsdale Healthcare Osborn Campus, Scottsdale, AZ, January 9, 2014.

7. Panjabi M, Dvorak J, Duranceau J, Yamamoto I, Gerber M, Rauschning W, et al; Three-dimensional movements of the upper cervical spine. *Spine* 1988, 3:726–730.

8. Boszczyk BM, Littlewood AP, Putz R; A geometrical model of vertical translation and alar ligament tension in atlanto-axial rotation. *European Spine Journal* 2012, 21(8):1575–1579. doi: 10.1007/s00586-012-2209-z.

9. WHO Guidelines to Spinal Manipulative Therapy. Specific Contraindications. http://wikichiro.org/en/index.php?title=WHO_Guidelines_-_Contraindications_to_SMT (accessed January 1, 2019).

10. Akinbo SR, Noronha CC, Okanlawon AO, Danesi MA; Effects of different cervical traction weights on neck pain and mobility. *Nigerian Postgraduate Medical Journal* 2006, 13(3):230–235.

11. Rammel ML; Relationship between therapist body weight and manual traction force on the cervical spine. *Journal of Orthopaedic and Sports Physical Therapy* 1989, 10(10):408–411.

12. Yee D; Mass of a human head. *The Physics Factbook.* https://hypertextbook.com/facts/2006/Dmitriy Gekhman.shtml (accessed November 29, 2018).

Index

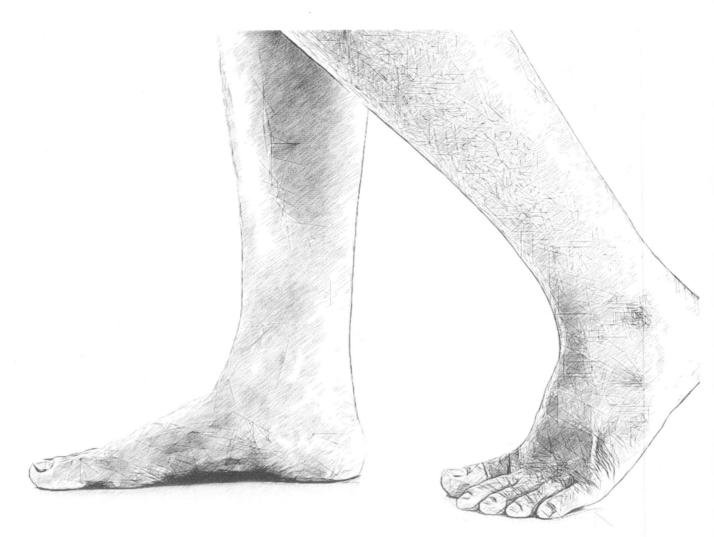

Take your next step!

Learn to see, apply and develop the content of the book. Join our education! Read more at asperaeducation.se

SOMA MOVE®

"*My heart is pounding hard
and the sweat is running down my face,
yet my breath is not out of control.
I breathe in harmony with every movement
I make and I can feel myself smiling.
Each fiber in my very being is alive and active.
I move over the floor in what seems to be a
neverending flow of movement.
I close my eyes and I feel completely focused,
I am here, I am now and*

I am movement."

s o m a m o v e . c o m

Visit our website!

Please visit our website to connect
with the authors, buy the posters
and our online courses.

movementintegrationbook.com